THE FIRST GREAT AIR WAR

Just eleven years after the Wright brothers' first flight, the Royal Flying Corps set off for France, and every aspect of air-fighting had to be discovered for the first time. The pilots and observers of all the warring nations tended to be young, courageous individualists, filled with a love of flying and of adventure. Their aircraft were fragile and unreliable and casualties were appallingly heavy. It was a period of great technical advances when aerial warfare and tactical teamwork were transformed.

THE FIRST GREAT AIR WAR is the story of air operations during World War I, of the careers and exploits of the fighter 'aces', of the less famous pilots, of the air-observers and gunners and of the less glamorous but vitally important crews who carried out reconnaissance and bombing sorties. It is a study of the formation and development of the Royal Flying Corps, and of the lives and experiences of the airmen who were its pioneers.

About the author

R. L. T. Bickers spent his early childhood in the 1920s in India, and in Iraq where he often visited the R.A.F. station near Bagdad and met many serving officers who had flown in the R.F.C. (later the R.A.F.) in the First World War. He holds a Permanent Commission in the R.A.F., in which he spent nearly twenty years. Joining at the outbreak of the Second World War, he operated with Fighter and Coastal Commands in England, North Africa and Italy. As he speaks and writes numerous languages, he is able to do all his own research. He has written many novels, also radio plays and documentaries, and a well received biography of 'Ginger' Lacey, the top-scoring pilot in the Battle of Britain.

The First Great Air War

Richard
Townshend Bickers

CORONET BOOKS
Hodder and Stoughton

British Library C.I.P.

Bickers, Richard Townshend,
1917–
    The first great air war.
    1. World War I. Western Front.
    Air operations
    I. Title
    940.4'4

ISBN 0-340-50824-8

Printed and bound in Great Britain
for Hodder and Stoughton
paperbacks, a division of Hodder and
Stoughton Ltd., Mill Road, Dunton
Green, Sevenoaks, Kent TN13 2YA.
(Editorial Office, 47 Bedford Square,
London WC18 3DP) by Richard
Clay Ltd, Bungay, Suffolk.

Acknowledgments

I thank the following for the essential facilities and information they provided during my research.

The Royal Air Force Museum
The Imperial War Museum
The Air Historical Branch, Royal Air Force, Ministry of Defence
Public Record Office
Australian War Memorial, Canberra
Public Archives Canada, Ottawa
National Defence Headquarters, Ottawa
South African Air Force Museum, Swartkop
Major General K. R. van der Spuy, SAAF Retd
Professor (ex-Lieutenant Colonel, SAAF) Vivian Voss
Royal New Zealand Air Force Museum
Service Historique de l'Armée de l'Air, Vincennes
Musée de l'Air et de l'Espace, Le Bourget
Ufficio Storico, Stato Maggiore Aeronautica, Rome
Militärgeschtliches Forschungsamt Abteilung, Freiburg im Breslau
Militärarchiv, Bundesarchiv, Friburg i Br
Wehrbereichskommando III, Zentral Bibliotek der Bundeswehr, Düsseldorf
Copies of the Albert Ball letters in the Nottinghamshire Archives Office are reproduced by permission of the Principal Archivist

Contents

Illustrations

Picture acknowledgments

1. By courtesy of The Royal Air Force Museum, Hendon.
2. By courtesy of Mr Bruce Robertson
3. By courtesy of the Imperial War Museum
4. By courtesy Service Historique de l'Armée de l'Air
5. By courtesy of Stato Maggiore dell' Aeronautica Militare, Ufficio Storico
6. By courtesy of the Robert Hunt Library

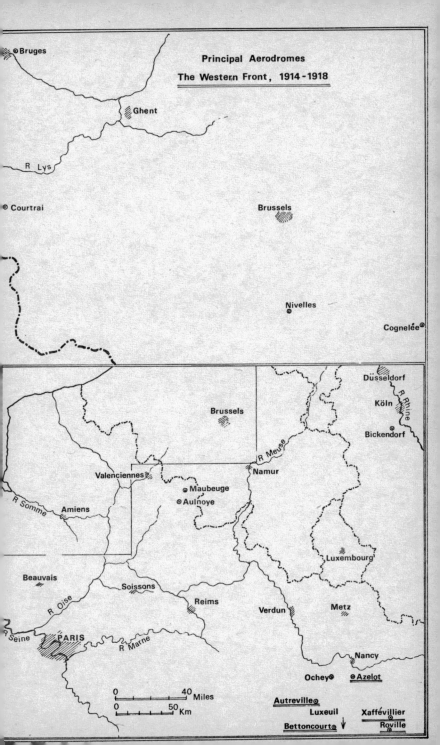

Principal Aerodromes

The Western Front, 1914-1918

Foreword

I do not remember the Great War, now called the First World War, but its reverberations are among my oldest memories.

In early childhood in the 1920s, when Iraq was under British mandate, I lived in Baghdad. Our expatriate community was small and the majority of my parents' friends were serving officers who had flown in the Royal Flying Corps, later the Royal Air Force, in 1914–18. The aerodrome at Hinaidi became almost as familiar to me as our own garden. There, I saw aeroplanes that had won fame on every Front, going about their peacetime duties.

I gazed up at them flying over our house. I waved excitedly in response whenever one made a jovial dive across our roof; and once, to my ecstasy, a message pouch attached to a red, yellow and blue streamer was dropped; as though it were a despatch for some infantry company beleaguered in a Flanders salient. Enviously, I watched the fighters at air displays: formation flying, aerobatics (still known as "stunting"), mock combat; and the bombers making dummy attacks on tanks lumbering across the desert. There were Bristol Fighters, Sopwith Snipes and Camels, de Havilland DH9As and the giant Handley Page 0/400 with its hundred-foot wingspan.

People often referred to the RAF as "the RFC", from old habit. Domestic entertainment in the outposts of empire was still Edwardian, late Victorian, even. Host, hostess and guests provided it musically for themselves. After a dinner party, if there was an impromptu concert around our drawing-room piano, my mother at the keys and leading with her trained contralto, my father adding his fine baritone and other ladies and gentlemen the whole range from soprano to bass, I used to hear, as I lay awake, that "Old King Cole was a merry old soul", who "called for his fiddlers three", and there was "none so fair as can compare with the men of the RFC": "RAF" would not have rhymed.

Successive Air Officers Commanding were Air Vice Marshals Sir John Salmond and Sir Robert Brooke-Popham – both later Marshals of the Royal Air Force – who, as junior officers, appear in the following pages. Small boys of my generation were taught to shake hands with adults, so I have felt the grip of distinguished hands that had shifted the joysticks of the earliest Service aeroplanes and fired Lewis guns at the enemy.

When I joined, in the next war, many of those with whom I served had been 1914–18 pilots and observers, and some were flying still. Coaxed to reminisce about their war time experiences in France, Mesopotamia or Italy, they enthralled me. The majority, of course, were workaday aircrew who wore no decorations: although they had survived many dangers, shot down their fair share of the enemy, repeatedly dropped bombs on heavily defended targets, strafed trenches at nought feet in the face of withering machinegun and rifle fire; exactly as their successors were doing in 1939–45.

But there were also prodigies whose feats I had admired since boyhood: one of whom, Louis Strange, DSO, MC, DFC, added a bar to the last of these in 1940, flying a Hurricane; and was one of the most electrifying personalities I ever knew: only Douglas Bader made the same instant forceful impact. And many of our highest rankers were such as Sholto Douglas, Collishaw and Coningham, who had been dazzling in youth on their squadrons and were brilliant now in command of great air forces.

Lord Chesterfield, writing to his son in 1748, declared "... there never were, since the creation of the world, two cases exactly parallel; and ... there never was a case stated, or even known, by any historian, with every one of its circumstances; which, however, ought to be known in order to be reasoned from.... Take into your consideration, if you please, cases seemingly analogous; but take them as helps only, not as guides."

His Lordship had never seen action. If he had, he would have known that there are many common factors between one war and another: the appalling din, which stunned soldiers at Waterloo, the Somme and Alamein, sailors at Trafalgar and Jutland, airmen among bursting anti-aircraft shells and the noise of their own guns in an air fight, ground strafe or shipping strike; the reek of cordite; the sizzling streaks of tracer bullets, whether on land, at sea or in the air; the stress and fear, and grief at the loss of comrades. Above all, for airmen, in 1914–18 as in

1939–45, and every subsequent war, is the horror of fire.

The threat of being burned alive is not unique to aircrew. Tank crews know it and so do ships' companies. Infantry are menaced by flamethrowers. Civilians are roasted by incendiary bombs. But there are some dangers and some brands of suffering, some attitudes towards the enemy and the job one has to do, to situations, experiences and reactions to these, that are peculiar to those who defy the law of gravity when they go to battle. The affinity between the First and Second World Wars is, in that respect, close.

The First World War had run two-thirds of its course by the time I was born, but I have had the good fortune to know many brave men who fought in it and to see, in flight and simulated action, some of the aircraft they flew. I have always felt that, although I was not there, I have a fairly accurate idea of what it was like; and of the attitudes and characteristics of those who pioneered aerial warfare. It helps to understand those times and those people, if one has some experience of what war in the air is like and of escaping from a burning aircraft. In my teens I went sometimes to the flying club at Brooklands, the cradle of British aviation. In the late 1940s I knew Netheravon, where the officers' mess was the same building where Mannock, McCudden, Ball, Rhys-Davids, Hawker and most of the other heroes of the RFC and its offspring, the RAF, had found recreation and rest after strenuous hours of flying the recalcitrant, dangerous aeroplanes of their day. One felt their presence.

CHAPTER 1

Going to War

On a fine spring morning, 1st April 1915, a French pilot flying a two-seater Morane-Saulnier L over Flanders fired seventy-two shots at a two-seater German Albatros and radically transformed the whole lineament of air combat.

The Great War was eight months old. The Morane had a Hotchkiss machinegun which, for the first time in history, could fire straight ahead between the blades of a spinning propeller. The pilot of the Albatros was armed with an automatic pistol and his observer with a rifle. The Morane was alone. The Albatros was in a formation of four.

When Lieutenant Roland Garros had emptied three magazines of ammunition there was no need to reload and shoot again. One Albatros was in flames and spinning earthwards. No parachutes were carried then: its occupants, if alive, were trapped.

The other three Albatroses, their crews aghast at this astounding new phenomenon of a front-engined aeroplane that fired a machinegun through the propeller, and bemused by the swiftness of the killing they had witnessed, did not tarry. They dived as fast and steeply as they could, heading for their aerodrome to report this scourge of the skies that had burst upon the Western Front.

Suddenly the first eight months of war were relegated to a past that would henceforth seem to have been Arcadian, almost playful, by comparison. At one stroke the life expectancy of every man flying on any battle front henceforth was shortened many-fold.

Garros was as shocked as the enemy by the cataclysmic spectacle. When he went to see the wreckage the revulsion he felt was a reaction that would be shared by thousands of other victorious pilots in that and every future war.

In his own words: "It was tragic, frightful. At the end of perhaps twenty-five seconds, which seemed long, of falling, the machine dashed into the ground in a great cloud of smoke. I went by car to see the wreck. Those first on the scene had pilfered souvenirs: sidearms, insignia and the like. I took energetic steps to retrieve them. The two corpses were in a horrible state, naked and bloody. The observer had been shot through the head. The pilot was too horribly mutilated to be examined. The remains of the aeroplane were pierced everywhere with bullet holes."

This repugnance at what he had had to do in the line of duty has been expressed by many pilots who have shot down enemy aircraft. It did not noticeably deter any from repeating the performance; nor should it have. There was no hypocrisy there. What they had done was unavoidable. The regret they admitted was matter-of-fact and unsentimental; but it was a sign of the instinctive respect which fighting airmen of all nationalities have shown each other from the outset.

Garros scored his second victory on 15th April and his third on the morning of the 18th. That afternoon he himself was shot down by ground fire near Courtrai and taken prisoner. His aircraft fell into the hands of the enemy and the secret of how he was able to fire a machinegun through the propeller was revealed.

After Garros had proved the efficacy of his invention, the French Military Air Service began to modify its tractor aeroplanes to fire through the airscrew, or mounted a machinegun on the upper wing to fire outside the propeller disc. The British emulated them, but slowly. The Germans adopted and improved on Garros's device. On 20th May, less than five weeks from the day Roland Garros was shot down, there were two Fokkers at the Front equipped with both an interruptor gear that synchronised propeller revolutions with rate of fire, and deflector wedges on the airscrew blades for those rounds which did not pass cleanly between them.

This radical innovation and the developments it provoked, however, were still in the future when the Royal Flying Corps set off to war. All that preceded it will appear in chronological order. Meanwhile, to put the arrival of the RFC at the Western Front in August 1914 in perspective, the sequence of events to which the present Royal Air Force owes its origin must be set in order.

When Germany declared war on France on Sunday 4th August 1914, her alleged justification, that an aeroplane of the Aviation Militaire had bombed the railway near Nuremberg and Karlsruhe, was more than a lie; it was the harbinger of a new weapon, a fresh element, a third dimension in warfare.

The Germany Army invaded Belgium, which Britain, France's ally, had promised to support. An hour before midnight the British Empire went to war. The British Expeditionary Force prepared to cross the English Channel and embarked six days later.

The Royal Flying Corps made ready to send all four of its squadrons to the Front, with the Headquarters Unit, and an Aircraft Park that held spare aeroplanes and parts. Sixty-three aeroplanes, rear-engined Henry Farman F20s with a pusher propeller, and front-engined Blériot XIs, BE2s, Avro 504s, and BE8s with a tractor airscrew, which took off from Dover on the 13th and 15th, assembled by the 17th at Maubeuge.

They appeared out of the summer sky, engines clattering and stinking of castor oil: flimsy structures of wood, canvas and bracing wires; many of them primitive-looking contraptions with a naked fuselage of ribs and spars.

They had flown across the Channel from Dover to fight the first great air war in history. They landed at varying intervals, according to how well their engines had functioned, how much the wind had affected them, how many forced landings they had made and how accurate their pilots' navigation. One had crashed in England and killed its pilot and his mechanic.

France confronted the enemy with twenty-five escadrilles, based at aerodromes across the country from Ostend to Nancy, equipped with a total of 142 aeroplanes: rear-engined two-seater Farman Longhorns, Voisins, Caudrons and Farman F20s; front-engined, single-seater Blériots and Morane-Saulniers. The twenty-one two-seater escadrilles each comprised six machines and were for general Army co-operation. Each of the four single-seater escadrilles, which operated with the cavalry, had four machines.

Germany sent thirty-three Field Flying Units to the Front. Each consisted of six tractor types. The Taube and Fokker were single-seater monoplanes. The majority, Albatros, LVG, Aviatik, AEG, were two-seater biplanes.

The disparities between the air forces were typical of divergent national characteristics, but all had recognised reconnaissance as the first function of an air Service: an extension of the cavalry's business. The British and French had also practised artillery spotting, and one type of aeroplane might have served both purposes. All three countries sought at least to provide for their needs with the smallest possible variety of aeroplanes. The sheer inventiveness of designers, however, had foisted a plethora of choices on both Britain and France. Each also had a military aircraft factory as well as private factories building aeroplanes for the approval of the RFC and l'Aviation Militaire. Naturally the Royal Aircraft Factory and le Service des Fabrications de l'Aéronautique were resentful if their products were not chosen. The Germans had left all their requirements in the hands of civilian manufacturers, who also provided all flying instruction.

The British had had little success in encouraging the indigenous manufacture of aero engines. Wolseley, Beardmore and Rolls-Royce were the leading makers, but in quantities too small to meet demand; and only the last-named was reliable. The RFC therefore had to depend on the French for Le Rhône, Clerget and Renault engines. The needs of l'Aviation Militaire obviously had priority. In consequence, deliveries to the RFC were slow and the engines often second-hand, reconditioned. Germany enjoyed high-volume production of excellent Mercedes and Benz. The nature of the power units strongly affected the design of airframes. The French built small rotary air-cooled ones, which meant that their aeroplanes had to be light. The Germans' were heavy in-line water-cooled and suited heavy aeroplanes.

The Germans had given some attention to mounting machineguns on their machines; and, although the British and French had been experimenting since 1913, neither had made much progress.

The three air contingents that converged on the Western Front were the precursors of the huge air fleets that would be familiar over Europe a quarter-century in the future.

But men, not machines, win or lose battles. Those who brought these first frail military contraptions to war were equally the forerunners of a new tradition, a mystique: a fraternity who met their opponents in an entirely new kind of combat and formed with them an empathy, shared a gallantry, that transcended national animosity. They created their own personal brand of conflict, with its private ethos and mores. From the beginning of time men had shown compassion and chivalry in battle on land and sea. This new breed, airmen, created a unique brand of good manners in mortal conflict and of wry, sardonic, understated fortitude that was to distinguish their successors in every war.

From whatever country had bred them, on whichever side they fought, they brought to their basic task common qualities of character: they flew for the sheer love of flying; they relished adventure; they enjoyed risks, and testing themselves and their aircraft to their limits. They were bold and adventurous far above the average.

Their special characteristics were evinced when Wilbur and Orville Wright took to the air. It imbued the fighter pilots in the Battle of

Britain; the bomber crews, of whom only one-third survived a tour of thirty operations; the Coastal Command torpedo strike crews, who had a mere seventeen and a half per cent chance of completing their first tour and three per cent of surviving a second. *That* was the significance of those few aeroplanes' appearance at the Western Front in 1914.

The RFC's arrival was the fruition of only twenty-seven months' planning, organisation and growth since its birth on 13th May 1912. It was a mere ten years and eight months after man's first flight in a heavier-than-air flying machine, which had lasted three and a half seconds, covered barely a hundred feet of ground and risen just fifteen feet above it.

This first overseas campaign by an air force might reasonably be likened to sending an army abroad with rifles, machineguns and shell-firing artillery less than eleven years after the invention of gunpowder; or a fleet of destroyers and cruisers to sea as short a time after the launching of the earliest sailing vessel.

CHAPTER 2

———◆———

Origins

In France, in 1783, Joseph and Jacques Montgolfier ascended in the world's first hot-air balloon and covered a mile and a half. The next year, their first hydrogen balloon rose from the centre of Paris and descended fifteen miles deep in the country. On its next sortie, it climbed to 1500 feet.

Ballooning became a craze which soon spread throughout Europe. It was an Italian, Vincenzo Lunardi, secretary to the Ambassador of the Kingdom of Naples at the Court of St James, who made the first aerial voyage in Britain, when he took a hydrogen balloon of 32-foot diameter up from the Artillery Grounds in London's Moorgate and landed some twenty miles away, near Ware. The date was 15th September 1784.*

Military aviation began with balloons. It was a paradox that such romantic, slightly frivolous, quasi-scientific objects, which provided a hobby for the unorthodox and were by custom gaily coloured, often garish, should be put to so serious a purpose. There was an aura of the fairground about them, not the battlefield, of jollity, not aggression.

Aviators, too, from the first, were distinguished by their high spirits and informality, and a special kind of eccentricity: a high degree of curiosity, and boldness in abundance, tinged with audacity. Anyone who took to the air had to be different in nature from the vast humdrum prudent majority of mankind in his craving for excitement; had to have an enquiring mind; had to be more than courageous: rash to the point of thinking it well worth while to break a limb or even his neck for the

* The same day, 156 years later, is now celebrated as Battle of Britain Day.

pleasure he enjoyed from being airborne.

The modern fighter and bomber, supersonically fast, formidably armed, capable of climbing from earth into the stratosphere in the time it takes a Wimbledon champion to win a love game, are the direct descendants of those sedately moving spheres that first drifted across the sky: lifted there by hot air or gas, driven only by the wind and at the mercy of its whims, incapable of being steered. Their pilots and crews are the progeny of those who cheerfully took off suspended in baskets beneath the huge bags of air or gas, not knowing to which point of the compass they would travel, how many involuntary changes of direction they would make, or, most daunting of all to normally conventional beings, where they would come down.

The airmen and aircraft of the Great War were the first modern links in the chain of evolution from the inception of aerial mobility to its culmination in space flight. Like all pioneers, the pilots and observers of that era had to be endowed with an even greater intellectual curiosity, love of adventure, and willingness to take risks that had a substantial prospect of proving fatal, than those who succeeded them. Balloonists were, in the British armed Services, soon known as "balloonatics": the ground-grippers looked on them – actually, stared up at them; probably with envy, wishing they had the courage to emulate them – as lunatics let loose with dangerous playthings. Two generations later, the adjective most often applied to Royal Flying Corps pilots was "wild"; and that also was used in admiration, and wistfulness at being excluded from what had quickly become an exceptional company, rather than pejoratively.

In Britain, the first Army balloon unit was formed by the Royal Engineers in 1878. By the next year it had a few trained officers and men, with, the archives state, "five reliable balloons". There is a pleasantly optimistic, Corinthian note about that "reliable", which makes one wonder just what vagaries and frustrations of performance and quality had to be put up with. That Easter, its Commanding Officer, Captain H. Elsdale, took Captain J. L. B. Templer, with a squad of troops and a balloon, to the Volunteer Review and manoeuvres in Sussex; and astonished the inspecting General as much as the spectators by "marching past" – the first recorded fly-past – at 250 feet; thus distracting all eyes from the columns of earthbound horse, foot and guns plodding past the saluting base.

In 1880 and 1882 reconnoitring balloons took part in the Aldershot manoeuvres. In 1884, Elsdale, now a major, led another officer with fifteen rank and file and three balloons to Bechuanaland as part of General Sir Charles Warren's expedition, but saw no fighting. However, the following year Major Templer went to Sudan with a couple of balloons and nine men, where he made some useful reconnaissances.

Artillery spotting from the air with tethered balloons was now tried. When the gas that inflated them deteriorated so much that they could not rise high enough for captive observation, training in free ballooning continued. Major Elsdale also experimented with aerial photography, using small free balloons carrying a camera operated by clockwork to expose the plates, then to destroy the balloon, so that it fell back to earth. Here again were the elements of what became, during the Great War, a sophisticated technique.

Italy formed an Army Aeronautical Section in 1884, and used its balloons during the Eritrean War of 1887–88.

In Germany, military ballooning was being tested on a small scale and with avoidance of publicity. Progress in France provided an incentive which soon led further: to fervent German interest in airships.

1889 was a decisive year for the eventual formation of a British military air arm. Lieutenants G. E. Phillips and C. G. Close, with a small detachment of Sappers, took part in that year's manoeuvres, at which the German Emperor was present. They operated with a force that set out to attack the enemy camp, and were asked whether there were any outposts to the rear of this. Balloon observation at dusk reported that the enemy was posting pickets only on the nearer side. A victorious night attack planned on this intelligence so impressed General Sir Evelyn Wood that an Army balloon section was formally created within the Royal Engineers the following year.

At the conclusion of the 1889 manoeuvres, Lieutenant Ward emulated Captain Elsdale eleven years earlier, with equally astonishing effect on the spectators: in the parade past the Kaiser, he was also towed at 300 feet in the basket of a balloon tethered to its wagon.

The South African War of 1899–1902 gave the balloonists the first protracted and most important opportunity to justify themselves. Hitherto, despite their demonstrations of tactical usefulness, they had been regarded rather as an exclusive club that demanded of its candidates certain standards for admission: predominantly, a touch of lunacy and a disregard for conformity, which promised to jeopardise their careers. Conventional soldiers looked on them with bemused ignorance. Many senior officers were prejudiced against them simply because their trade imported an experimental element into military operations. Innovations aroused suspicion in the often obtuse brains of field commanders; who were traditionally engaged in perpetual internecine strife to protect their own interests. The high-ranking officers in the War Office and their counterparts at the Admiralty were each perpetually scheming to obtain the largest share of the annual Defence Budget: while within the Army, those who commanded the artillery, infantry, cavalry, engineers, ordnance, supply and transport were as bitterly engaged in promoting the claims of their own arms. When confronted with a startlingly new concept, such as balloon reconnaissance and observation, it was easiest to condemn it, untried.

To justify scorn or mistrust of the newfangled ancillary, there were many legitimate faults to find. As is not uncommon with any novelty, other arms had not been properly trained in co-operation. The method of signalling by flag between ground and air was inefficient. The balloon observers were often unable to engage the attention of the gunners for whom they were spotting. The hydrogen for inflating the balloons was in steel containers that were heavy and bulky.

Despite prejudice, lack of preparation and difficulties in handling and transporting the equipment, balloons proved their worth. At the Battle of Lombard's Kop they reported the enemy positions and directed artillery fire. At Magersfontein they ranged the howitzers onto Boer cavalry in a gully concealed from viewers on the ground. At Paardeberg they reported General Cronje's dispositions, which enabled these to be attacked successfully.

In 1894, Farnborough, now famous in the history of aviation, entered the scene as the site of the Army balloon factory. The Superintendent, Templer, by then a colonel, was succeeded in 1906 by Colonel Capper, a prophetic protagonist of aeroplanes, who, lecturing at the Royal United Service Institution, said:

"In a few years we may expect to see men moving swiftly through the air ... such machines will move very rapidly ... up to a hundred miles per hour ... they will be small and difficult to hit ... their range will be very large."

Balloons were about to be superseded: in 1907, the British Army's first airship, 120 feet long, made a three-and-a-half-hour flight from Farnborough and circled London.

"It is the work of a lunatic," exclaimed one of the admirals who set eyes on the Royal Navy's first airship, the 512-foot *Mayfly*. He might well have made the comment whatever her size and performance, in reactionary revulsion from any vessel that presumed to navigate the sky rather than the sea. But in this instance there was some justification for his obloquy. Size she had in ample measure; but performance, literally none: except, embarrassingly, as an unintentional tragi-comic turn.

The gigantic aircraft had been under construction for two years. Interest, in the Service and among the general public, was intense, expectations were high. But, calamitously, she never flew. Brought out of her hangar in May 1911, she spent four days undergoing mooring trials tethered to her mast. It was found that her buoyancy did not sufficiently exceed her weight to give her the lift she needed to rise further aloft. So she was taken into her shed again and modified. In September, she emerged; and promptly broke her back. The wreckage was incapable of reassembly. Thus ended the first attempt to provide the fleet with a vantage point higher than a crow's nest.

The road that led to this disaster had been a short one since, in 1908, the Director of Naval Ordnance had proposed to the First Sea Lord that Vickers, Son and Maxim, who had satisfactorily designed and manufactured submarines, should be asked to build a rigid dirigible similar to the Zeppelin. Here was Naval loyalty, feeling for tradition, and general conservatism finding typical expression in its own peculiar form of logic.

Characteristically, the Navy and the Army had gone their separate ways in developing an eye in the sky. While the Navy had turned to private industry, the Army had relied on His Majesty's Balloon Factory at Farnborough. The two Services' requirements also differed. The land forces needed balloons and aeroplanes, and had flirted with airships. At sea, balloons would have been useless, but both aeroplanes and airships were necessary. These last enjoyed several advantages. They had far the greater range. They could carry the heavier load, so wireless equipment that sent and received across hundreds of miles was installed. They could stop engines and hover, to detect mines and submarines. Their crews could be accommodated for long periods in enough comfort to ensure decent rest and the maintenance of efficiency.

But, in the Army at least, the day of the aeroplane was at hand. Captain Bertram Dickson, Royal Field Artillery, wrote a far-sighted and liberal-minded memorandum to the technical sub-committee created by the Committee of Imperial Defence to advise on measures that would ensure Britain an efficient aerial Service:

"In the case of a European war between two countries, both sides would be equipped with large corps of aeroplanes, each trying to obtain information of the other, and to hide its own movements. The efforts which each would exert in order to hinder or prevent the enemy from obtaining information would lead to the inevitable result of a war in the air, for the supremacy of the air, by armed aeroplanes against each other. This fight for the supremacy of the air in future wars will be of the first and greatest importance, and when it has been won the land and sea forces of the loser will be at such a disadvantage that the war will certainly have to terminate at a much smaller loss in men and money to both sides." Wise sentiments and an advanced concept. The Captain was, it seems, also an optimist of no small dimensions; to judge from his concluding sentence.

The first airship, like the first balloon, was also French. Forty-four feet long, with a three horsepower steam engine, it flew on 24th September 1852 at six miles an hour.

All over the Continent, progress in the design of dirigibles was slow. After many vicissitudes, the first, still non-rigid, was *La France*, built for the Army in 1884. Shaped like a fish, 165 feet long, it was driven by an $8\frac{1}{2}$ h.p. electric motor.

Its successful flights prompted Germany to surpass the rest of the world, and in 1897 the first rigid airship took off, its 12-h.p. Daimler engine driving twin airscrews. In 1900, Lieutenant Colonel Count Ferdinand von Zeppelin flew the first of the huge dirigibles that were to bear his name. It was 420 feet long and had two 16 h.p. Daimlers in tandem, each driving two airscrews. In 1907, he made an eight-hour 211-mile flight in his third Zeppelin. By 1914 Zeppelins were making even longer and trouble-free flights.

The evolution of dirigibles from 120-foot non-rigid to rigid craft of over 400 feet, and through various shapes, was accompanied by frequent wreckage and much loss of life. Some caught fire while airborne, others burned out on the ground. Each country had its failures and tragedies and the German experience is typical of all. Of twenty-five Zeppelins built between 1900 and 1914, nine were wrecked while going about their peaceful occasions. Three were destroyed by fire. In the five months following the declaration of war, four were shot down by anti-aircraft guns and one wrecked itself.

Clearly, it was on the aeroplane that development must concentrate; and it was in the United States that the swiftest and furthest progress was expected, since it was there that Wilbur and Orville Wright had flown the world's first aeroplane, on 17th December 1903. But once more it was France that forged ahead of the rest of the world.

That aviation evolved at all beyond the balloon is in itself remarkable, in view of the hazards that attended its practitioners on every hand. As if the disappointment – often felt as the disgrace – of failure, the threat of cataclysmic accidents and mid-air conflagration, the permanent shadow of death or severe injury hovering over every enterprise, were not discouragement enough, to all these was added the violent hostility of onlookers when some technical defect or meteorological adversity deprived them of the spectacle for which they had assembled; always in their hundreds, often by the thousand. In the USA, France, Germany, and Britain, in Spain and Scandinavia and Latin America, crowds fell upon aeronauts who failed to provide them with a sight of the wonders they had come to see. Looking back on it several decades later, the behaviour of disappointed spectators has its humorous side, despite their assaults on blameless airmen and mechanics.

In 1910, a British Army pilot, Lieutenant Lancelot Gibbs, who had trained in France, was about to give a flying demonstration at Durango in northern Spain. He kept the 30,000 crowd waiting more than an hour while his Farman biplane, which had come crated, by train, was being assembled. The mob became restive. To placate them, it was wheeled into the open so that they could see the work being done. This had the opposite effect. They damaged it so badly that it had to be taken back under shelter: whereupon they hurled stones, smashed the shed and disabled the mechanic. Someone pulled a knife on Gibbs, yelling that it was impossible for man to fly. This inflamed the rest of the thirty thousand, who screamed "Down with science! Long live religion!" and burned down the shed and the aeroplane in it, while the police escorted Gibbs – who had remained cool throughout – to safety.

The first aeroplane flight in France was made by a Brazilian expatriate, Alberto Santos Dumont, when he flew eighty metres on 23rd October 1906. Among those who competed with him were some whose names were soon illustrious in aeronautical history: Henry Farman, who lived his whole life in France and spoke little English, but had a British father; and native Frenchmen such as Blériot and the brothers Voisin. On 13th January 1908, Farman won a prize of £2000 for being the first in France to fly one kilometre around a closed circuit. In that year, Wilbur Wright visited France. His many flights, including the demonstration of figures-of-eight, helped to educate his rivals as well as breaking the national records for endurance, distance and height. He finally covered seventy-seven miles in two hours and twenty minutes.

In America, in May 1909, Glenn H Curtiss flew 120 miles in two hours and a half. On 25th July of the same year, Louis Blériot crossed the English Channel in forty minutes. In Germany, while the design and manufacture of airships drew worldwide attention, secrecy covered the work being done on aeroplanes in the Ago, Albatros, DFW, Halberstadt, Rumpler, and Harland factories, and the one belonging to the Dutch designer, Anthony Fokker.

On 8th June 1908, A. V. Roe made the first flight in England: of sixty yards at a height of two feet. Thus was the distinguished Avro marque founded; and soon T. O. M. Sopwith and Geoffrey de Havilland and the Short brothers were building aeroplanes and flying them. In those days, any Briton who aspired to being a pilot had either to teach himself or go to France to be taught. Flying caught on as swiftly as had motoring; although not, of course, on so wide a scale. The British, naturally, were quick to see the possibilities of a new form of sport in aviation, rather than its military potential or its contribution to science. Flying appealed largely to the type of man whose idea of bliss was to hurl himself down snowy mountains in the infant pastime of skiing, to hurtle through the bends of the Cresta Run on a skeleton toboggan, to drive at the highest speeds yet attained by man around a motor racetrack.

In 1909, the world's first international flying meeting was held at Reims. When one of the pilots who had been admired there, Monsieur Louis Paulhan, came to England to display his prowess, the Manager of Brooklands had the ground in the centre of the racing circuit cleared for his use. Paulhan's performance, culminating in a ninety-six-mile flight that lasted nearly three hours, so impressed the Brooklands Automobile Racing Club committee that a permanent aerodrome was laid out. In the following spring a busy community of aeroplane builders and pilots began to establish itself in a colony of wooden huts. A Mrs Hewlett, who had learned to fly, started a flying school there in partnership with a French pilot named Blondeau. By 1911 The Bluebird restaurant had become the social centre: taking, in fact, the place of that indispensable English institution, the local pub.

Brooklands attracted people from a great variety of occupations and incomes. Common interest bound them all in an easy sodality free from snobbery. Informality, friendliness and mutual help characterised it. Tools were readily lent. An outsider could not easily distinguish between the mechanics and those who employed them. Everyone wore overalls. In hot weather they preferred to work in pyjamas.

Hendon, in North London, had also become the site of a thriving aerodrome, where several manufacturers occupied the workshops and hangars and the public paid to watch displays. The first home of Army flying was established at Larkhill, on Salisbury Plain.

Air races began to be held, in Britain and abroad, and attracted entrants from many countries.

Meanwhile the establishment at Farnborough had changed its name to the Army Aircraft Factory and there, aeroplane design and manufacture were also going ahead.

Early in 1910, the Italian Army acquired its first aeroplanes, "a heterogeneous assortment under foreign labels," according to the archives, "Blériot, Etrich, Nieuport, and Farman biplanes. Powered by engines of 45 to 65 h.p., flying at 80–90 kilometres an hour, they succeeded in toiling up to more than 1000 metres with only the pilot aboard, and

500–600 metres when he was accompanied by an observer."

In 1911, when the Italo-Turkish war broke out in Libya (a Turkish colony since the sixteenth century and now coveted by Italy) a grandiloquently designated "Air Fleet", consisting of two Blériots, two Farmans, two Etrichs and three Nieuports, all with 50-h.p. motors, was taken by sea from Naples to an improvised aerodrome at Tripoli. The official historian notes that it was near the Jewish cemetery: which must have been more than a trifle disconcerting for the aviators; a macabre reminder of mortality, when every take-off in those primitive machines threatened to be fatal.

The honour of making the world's first operational sortie in an aeroplane fell to Captain Carlo Piazza, the Commanding Officer, flying a Blériot, on 23rd October 1911.

The majority of flights were on reconnaissance, "as had been foreseen", the archivist tells us: so the Italian Army had manifestly been as sharply foresighted as the French. "This employment proved opportune and rewarding." In addition, photographic sorties were flown; and, for the first time in history, the direction of artillery fire was done from an aeroplane.

The first air raid ever made is credited to Sub-Lieutenant Giulio Gavotti, who dropped one bomb on Ain Zara and three on Tagiura. "These bombs, named 'Cipelli', after their inventor, were spherical in shape, little bigger than an orange, and weighed two kilogrammes; they were dropped overboard by hand, after the pilot had wrenched off the safety catch with his teeth, to leave his hands free for flying the aeroplane."

"The exploits of the Italian airmen resounded around the world," the official records claim. Certainly *The Times* of 12th August 1912 admired them: "Nobody can have witnessed the deeds performed by the aeroplanes at Tripoli without feeling profoundly impressed by the courage and skill of the Italian pilots, and without being convinced of the practical value of aeroplanes in time of war."

Berliner Tageblatt on 10th September 1912 wrote: "The Italian Headquarters, thanks to its aeroplanes, is always informed about whatever movements are made by the Turkish troops, and, in consequence, always knows their disposition." It added: "The old maps of the French General Staff. were inexact and unserviceable; the map designed and printed by the Italians, based on photography, and the relief details obtained from the air, will be most useful for future operations."

There was no lack of predictions about the inevitable development of aerial warfare. The September 1912 issue of the British magazine *Central News* declared: "This war has clearly shown how the aeroplane, capable of penetrating deeply into enemy territory, constitutes in the near future a terrible method of destruction. This new instrument of war is destined to revolutionise modern strategy and tactics." So some journalist, evidently, was more perceptive than many senior British officers and Ministers. He went on to say: "What I have seen in the deserts of Tripolitania has convinced me that a great British air fleet needs to be created."

After the successes of the air force in Libya, Major Giulio Douhet, a respected visionary of aerial warfare, wrote: "A new weapon has appeared: the weapon of the air; a new fact has presented itself in the story of war: the principle of war in the air." The archivist's comment on this is: "Nevertheless – *nemo propheta in patria* – so modest was the development of the air arm in Italy, that on entering the war against Austria in 1915 our Army had only a few score aeroplanes, still of French construction or design."

CHAPTER 3

---◆---

The Build-up

It was difficult to persuade anyone in the British Services or Government of the importance that air power would have in the next war. Its advocates faced sheer incredulity as much as ingrained reactionary prejudice. None the less, the Air Battalion of the Royal Engineers was created on 1st April 1911, consisting of Headquarters and Number One Company, airships, at Farnborough; and Number Two Company, aeroplanes, with Captain Fulton commanding, at Larkhill. Officers were to be selected from any arm or branch of the Service for six months' training on probation, followed by four years' attachment. Apparently, they were not expected to devote the rest of their careers to aviation. Other ranks were to be taken only from the RE.

This was a time of some bewilderment about the direction that Army aviation should take. There was no emphasis on aeroplanes: they, airships and balloons were all within the warrant of the new formation. It was commanded by Lieutenant Colonel Sir Alexander Bannerman, who was not an aeroplane pilot, but a balloon specialist. But there were others who were outstanding protagonists of heavier-than-air machines. Captain J. D. B. Fulton, who had qualified as a pilot in 1909, and Captain Bertram Dickson, who "got his ticket" in 1910 were both Gunners. The latter died in 1913 from injuries in an aerial collision three years earlier. Captain Lancelot Gibbs was a third who shares with them the distinction of being the first British Army pilots. He soon had to abandon flying after an accident.

The Admiralty allowed Captain E. L. Gerrard, Royal Marines, and three naval officers, Lieutenants C. R. Samson, R. Gregory and A. M.

Longmore,* to be taught to fly. This was done at Eastchurch, which became the centre of Naval aviation.

Two significant developments occurred at the 1910 autumn manoeuvres. A wireless signal was transmitted from a Bristol Biplane to a receiver a quarter of a mile away; and Captain Dickson made an attempt to use another of these aircraft for scouting. The cavalry automatically objected on two grounds: that it would usurp their function, and, moreover, frighten their horses. One reconnaissance flight was agreed, however, but had to be cancelled on account of gusty winds; which did not enhance aviation in the eyes of the military in general.

Of the many misconceptions that survive about the air force in its earliest days, perhaps the most insistent concerns the type of man it sought as ideal pilot material. The dogma that good horsemen would make the best pilots, because they had "good hands", prompted interviewing officers to ask applicants "Do you ride?" This in turn led to the erroneous supposition in later years that the majority of those selected were cavalrymen. This was not so. To be a competent horseman was not synonymous with being a lancer, hussar or dragoon. In the Edwardian and early George V eras, to ride was a normal accomplishment of the upper and middle classes – "the officer caste" – for sport or as a means of transportation. Gentlemen Cadets at the Royal Military Academy, Woolwich, preparing to be Artillery and Engineer officers, and at the Royal Military College, Sandhurst, bound for the infantry, were put through riding school. Every commissioned officer was, therefore, a horseman. Of the 105 RFC officers who went to France in August 1914, a mere five had transferred from the cavalry and one from the Royal Horse Artillery. There were eleven other former artillerymen. All the rest had come from the infantry, except for one doctor and one member of the Intelligence Corps.

Both initial and continuation training were haphazard. The spirit of improvisation, adventure and joyous uncertainty prevalent at Brooklands pervaded Larkhill. The new Army pilots were instructed on a miscellany of Blériots, Farmans and Bristol Boxkites. Remarkably, there were no accidents. Civilian clothes were worn for flying. Cross-country work predominated and pilots spent about half their time away, delayed sometimes as long as a week by engine trouble; or after having forced-landed on some country estate where they were well entertained. No practices in co-operation with other arms were attempted.

In 1911, also, there appeared at the Bristol Flying School at Brooklands a figure destined to be of seminal importance in the story of what

*Later Air Chief Marshal Sir Arthur Longmore, GCB, DSO.

[17]

was to burgeon from the Air Battalion, bear its first blossoms as the Royal Flying Corps and attain full bloom as the Royal Air Force. He was Brigadier General David Henderson, a Scot, born in Glasgow on 11th August 1862, and known as the handsomest man in the British Army. An incident that occurred while he was at Staff College reveals his unassuming and modest nature. Needing to find the Mess Sergeant, he went to the latter's married quarters, instead, as most officers would, of sending a runner to fetch him. He happened to be wearing full dress. The sergeant's wife answered the door but stood staring at him, dumbstruck.

"What is the matter, Mrs X, are you ill?" Henderson asked.

"Lord, sir," she replied, "I can't take my eyes off you."

Her admiration merely amused him. Conceit, like pomposity, had no place in his character.

When he enrolled at the flying school he did so as "Mr Henry Davidson": to avoid the special attention he would otherwise have received. Living at Byfleet, a few miles from Brooklands, and working at the War Office, he had to fit in his daily flying lessons at dawn.

This dedication was not unexpected in a man whom John Buchan described as "The perfect combination of the two Scottish race stocks, the lowland and the highland, the Covenanting and Cavalier. He had the shrewd canniness of the lowlands, their long patience, their dislike of humbug, their sense of irony in life and character. And he had, too, something of the tough knuckle of obstinacy which goes with these endowments. A touch of the 'Shorter Catechism' was not wanting, for he had an austere sense of duty and a vigilant conscience. On the other hand were imagination and a warm generosity of heart. He was always extraordinarily susceptible to new ideas and quick to kindle. He had his countrymen's capacity for honest sentiment; tradition and romance played on his mind like music; and behind his reserve lay something gay and adventurous and debonair. All this might be read in his face, one of the handsomest I have ever seen."

Major General the Hon J.E.B. Seely (at that time, Lieutenant Colonel) described him as "a most remarkable man, with the qualities of courage, constancy, charm and industry, each in exceptional degree". Courage it certainly required to learn to fly at the age of forty-eight when everyone told him that, with the aeroplanes of that era, it was a most dangerous and indeed foolish enterprise.

Captain Howard Pixton, who taught Henderson to fly, wrote to a friend: "He stands out in my mind as being one of the finest and straightest men I have ever known. He had a beautiful 'touch' and would have made a splendid aviator had he gone in for it thoroughly. He picked it up straight away and had magnificent judgment. He was flying solo in about two or three days. Just about then there happened to be only two or three pupils and the weather was good, so that he got in a fair amount of practice. He was a born flyer."

In obtaining his "ticket", the Royal Aero Club certificate, Henderson set two records: he was the oldest pilot in the world, and had qualified in the shortest time: one week.

Air Chief Marshal Sir Robert Brooke-Popham, then a captain, said of him: "My first impression was of an exceedingly *kind* man full of vivacity and keenness." He wrote, also: "The fact that a senior officer working at the War Office had found time and had a sufficiently adventurous spirit to learn to fly, created a great impression among the few Army officers who at that time were intending to take up military aviation. I suppose most of us in those days had to withstand countless arguments from relations, personal and Service friends, in order to adhere to our intention of flying; not only was it looked upon as extremely dangerous, but the majority of people thought there was little in it from a Service point of view, and that to join the Air Battalion meant the end of a career that might otherwise have been successful."

It cost £75 to take the pilot's course, which had to be passed before acceptance into the Air Battalion. Despite the War Office's promise to refund this, none of the pupil pilots had received it. Henderson himself had to wait five months before Government honoured its commitment. The enterprising and amiable flying school overcame this major difficulty by taking pupils on credit: in the true spirit of the day, when all aviators, embryo and qualified, were brothers in what outsiders thought was foolhardiness.

When, that same year, Government appointed Colonel Seely to be Chairman of a new sub-committee to draw up a scheme for the development of the air force of the future, he consulted General Sir John French about the best man to represent the military side. The answer was: "Without doubt David Henderson, for two reasons: the first that he has learned to fly, a very rare thing nowadays; and I suppose an air sense to an airman is as important as sea sense to a seaman. Secondly, he is a faithful man. He will not fail you in a tight corner."

The other members of the committee were Major D. S. MacInnes and Captain F. H. Sykes.

The new arm was to take part officially in annual manoeuvres for the first time, that August, in Cambridgeshire; but severe drought forced a cancellation. However, the Aeroplane Company had already set off from Larkhill for Cambridge and perhaps its misadventures en route provided as valuable experience as the exercise would have. Engine failures, forced landings and crashes – in which nobody was seriously hurt – left only two aircraft to reach their destination. Navigation methods were less than refined and at least one officer had to rely on a map taken from Bradshaw's railway timetable.

In contrast with this off-hand performance, the French, although no separate air force yet existed, showed considerably greater professionalism and intelligence in co-operation with ground troops, also in August that year. By then, France had more than 200 military aeroplanes: Farman biplanes and a variety of monoplanes, Antoinettes, Blériots, Deperdussins and Nieuports, among them. Training was thorough and the air force frequently practised co-operation with artillery, cavalry and infantry. Captain R. Glyn of the Air Battalion, who witnessed these summer manoeuvres, reported to the British Government that the general employment of aeroplanes with troops had increased fighting efficiency by twenty per cent.

From the outset, the French had identified reconnaissance, and the observation and direction of artillery fire, as the most important uses of the air. For the former, close co-operation with cavalry was practised. For the latter, special gridded maps were used by which pilots or observers could notify, and battery commanders record, the fall of shells with an accuracy of a few metres. Infantry also were taught to make the best use of air co-operation, and aerial photography was initiated.

Secrecy in Germany continued to ensure that little was known about the use there of military aeroplanes or the number of her aeroplane pilots. The only certain information was that she had a strong force of airships which had the range to bomb any part of both France and Britain; and give early warning of enemy fleets approaching her shores, while they were still many hours' steaming away. In fact, the Army had thirty-seven aeroplanes and thirty pilots.

At least one German military genius, however, was giving his attention to heavier-than-air machines, and aerial defence as well as aggression: Ludendorff, later, as a general, to direct operations on the Western Front, was then a Staff colonel. In a paper submitted to the Chief of the General Staff, he recommended the establishment of ten service and six home defence units. The CGS felt that this would not be adequate to compete with France. He accordingly increased it to thirty-three and ten, respectively. He also insisted on experiments in night flying, hitherto not attempted. It was obvious, he observed, with considerable understatement, that attacks in the dark would make the enemy (by whom he meant the French) "uneasy". It was significant that he had, again, attack rather than defence in mind.

Some important advances had been made in 1911, but January 1912 found Britain with only eleven Army pilots (including Henderson, now approaching his fiftieth birthday) and eight in the Royal Navy.

France had 263.

A year after Henderson learned to fly, another impressive character – domineering, where Henderson was persuasive, bullying, where Henderson was gentle – Major Hugh Trenchard, decided that he, too, would acquire the Royal Aero Club's certificate. On 18th July 1912, within some seven months of his fortieth birthday, he became a pupil at T. O. M. Sopwith's Brooklands flying school. The War Office had promised him that if he passed, he could attend the course at the Central Flying School which was to start four weeks hence.

Mr Sopwith said of him: "It was no easy performance to undertake, but Major Trenchard tackled it with a wonderful spirit. He was out at dawn every morning, and only too keen to do anything to expedite tuition. He was a model pupil from whom many younger men should have taken a lead."

A paragon of ambition and enthusiasm he might have been, but the evidence is that he was an irascible martinet with it. A fellow pupil, Captain Edward Ellington,* who had replaced MacInnes on the aeronautical sub-committee, has recounted that early one morning when they were both waiting to start the day's work, Trenchard loudly cursed the weather and the unpunctuality of their instructor, Mr Perry. "Perhaps he has an explanation," Ellington suggested. "He'd better have a thundering good one," Trenchard growled. Perry's apology did not mollify Trenchard, who, to use a colloquialism of the RAF's thirty years later, tore him off a monumental strip.

The standard programme of training through which Perry put Trenchard is interesting more for what was omitted than for what was included. He began on 18th July, with two circuits in ten minutes, as passenger, watching the instructor's actions. On the 20th he did some more dual circuits, but at the controls. By the 26th he was ready to do two taxiing runs on a Farman ("ground rolls" they were called, a horrifying notion in modern flying jargon). He spent ten minutes in the morning and again in the afternoon of the 27th flying figures of eight. Weather prevented take-offs for the next three days. On 31st July he passed his tests, having flown a total of sixty-four minutes in thirteen days. On 13 August, he was given his certificate.

Another minor conflict, the Balkan war, flared up in 1912 when Bulgaria, Serbia, Greece and Montenegro formed an alliance to free Macedonia

*Later Marshal of the Royal Air Force Sir Edward Ellington, GCB, CMG, CBE.

from Turkish rule. The Bulgars were surprisingly air-minded. Having, however, few pilots or aircraft, they employed mercenaries who brought their own aeroplanes. Among these was the swashbuckling American Bert Hall, who was later notorious in the Great War for his boasting and the liberties he took with the truth. Hall knew that another American, Riley Scott, had invented a combined bomb sight and bomb rack, the previous year: the world's first and only instrument for either function. Hall had the device copied and was able to bomb with greater accuracy than his comrades, who sighted by guesswork; and attached a twenty-two-pound bomb to one foot by a slipknot, which they kicked free over the target. Another mercenary pilot, a Russian named Sakoff, was shot down and killed by rifle fire from the ground: the first airman ever killed in action.

This was also the year when France and Germany formed their air forces. The major decisions about the organisation of aeronautics in the British armed forces were resolved on 13th May 1912, when the Royal Flying Corps came into being with a Military Wing, a Naval Wing, and, at Upavon, a Central Flying School for the instruction of pilots from both.

When Trenchard arrived at Central Flying School from Brooklands, its Commandant, Captain Godfrey Paine, RN, appointed him Adjutant. It was already obvious that he would rise high in the new Service, although Longmore, his instructor, described him as: "At best an indifferent flyer ... his age told against him." But age had not handicapped Henderson, who was ten years older. Trenchard's height and heavy weight also made him an unwelcome passenger in the small low-powered contemporary aeroplanes.

A year later Henderson promoted him lieutenant colonel and second-in-command. He earned great esteem at Upavon by his drive and for his thorough familiarisation with every aspect of the school's work. He sought also to minimise accidents, in what was an inherently dangerous activity, by strict discipline and adherence to precise training programmes. As usual, he was respected, if not universally liked.

The Military Wing's function was defined at its inception: to operate with the other branches of the Army by carrying out reconnaissance to provide intelligence on enemy strength, disposition and movements. It would comprise 133 officers. Specially attractive terms would be offered to technical ground staff, in order to attract the most skilled. The war establishment called for 364 pilots, of whom half should be officers and half NCOs. Here was evidence of the comparative lack of social discrimination that was always to distinguish the RFC and is still notable in the RAF. No provision was made for specifically trained observers.

The plan for the Naval Wing, however, remained nebulous. It was still experimenting. There was not yet a clear design for the use of aircraft in Naval tactics. Only thirty or forty officers would initially be required.

The status of the new formation did not immediately warrant high ranks. Command of the Military Wing was given to Captain F. H. Sykes, with Lieutenant B. H. Barrington-Kennett, Grenadier Guards, as Adjutant. The latter's declared intention was to ensure that it would have the smartness of the Guards and the efficiency of the Royal Engineers. In time, both ideals were attained. By the end of the First World War, the RAF was – and has remained – the most efficient air force in the world; and the Colour Squadron of the RAF Regiment, which came into being in the Second, is renowned on parade.

The pride of the Senior Service and its assertion of the right to independence soon became manifest. Within a few months it ceased to call its air arm the Naval Wing of the Royal Flying Corps and unofficially substituted the title Royal Naval Air Service. There was then no need for the appellation "Military Wing" to continue in use: reference to the RFC henceforth became recognised as applicable only to the Army The designations "Airship Company" and "Aeroplane Company" also ceased to exist. Squadrons were born. They were to consist of three flights, each with eight pilots and four aircraft; plus the squadron commander and his.

The Airship Company, based at Farnborough, had descended from the Balloon Section of the Royal Engineers, was therefore the senior unit, and became No. 1 Squadron. Its Commanding Officer, Captain Edward Maitland, was one of the few pilots who still preferred airships. He had begun aviating with balloons in 1908, and that year, accompanied by Mr C. C. Turner and Professor A. E. Gaudron, made a thirty-six-and-a-half-hour journey of 1117 miles, which began at the Crystal Palace and ended in Russia, at Mateki Derevni. He progressed the following year to flying aeroplanes. In 1913 he made the first parachute descent from one. Two years later he parachuted from a balloon, at a height of 10,500 feet.

Captain C. J. Burke took command of No. 2 Squadron, of which the aeroplane pilots at Farnborough formed the nucleus. What had been the Aeroplane Company at Larkhill became No. 3 Squadron, under Captain H. R. M. Brooke-Popham.

More aeroplane squadrons were planned, using the flights of the first two as the nuclei for each. The establishment was raised to a major in command, with captains leading the flights.

Advances in technology were not reducing accidents. The Corps's first fatalities occurred on No. 3 Squadron in July, when Captain Loraine and his passenger, Staff Sergeant Wilson, crashed. In September, two officers were killed in a Deperdussin, on an exercise with the cavalry. Four days later, two more died in a Bristol. These three accidents prompted the War Office to prohibit monoplanes.

The young Service was looking for a biplane that was sturdy and fast, climbed well, could operate from rough ground and had a short landing run. The Royal Aircraft Factory's 70 m.p.h. BE2 met these requirements fairly well. In August, flown by Geoffrey de Havilland, with Major Sykes as passenger, it set a British altitude record of 10,560 feet, in 80 minutes, and was ordered as standard equipment for the squadrons.

In the following month, before deliveries could begin, Nos 2 and 3 Squadrons took part in the Army manoeuvres. The former operated with the two divisions constituting the attacking force, under General Haig, the latter, with the defending force of two divisions commanded by General Grierson. On the first afternoon of the exercise, the Cavalry Commander told Grierson that, on account of the distance separating the opposing forces, he would be unable to supply information about the enemy until two days later. Grierson referred the matter to Brooke-Popham, whose aircraft took off at 6 a.m. next day and returned three hours later with the detailed information that the Force Commander needed. Thereafter he planned his tactics entirely on aerial reconnaissance.

After the manoeuvres, a flight was detached from No. 2 Squadron as the foundation of No. 4 Squadron, commanded by Major G. H. Raleigh.

During the remaining three months of 1912, all three aeroplane squadrons increased their pilot and aircraft strength; and No. 3 Squadron moved from Larkhill to a new aerodrome nearby, Netheravon, where it began teaching non-commissioned officers to fly. The first to qualify was a rigger, Corporal (later Sergeant) F. Ridd. Another was W. T. M. McCudden, whose younger brother, James, joined the squadron as a mechanic in 1913, qualified as a pilot in 1915 and became one of the most respected squadron commanders in the Service, decorated with the VC, DSO and MC.

In August 1913 the RFC attained further independence when Brigadier General Sir David Henderson, already Director of Military Training, was appointed Director General of a new Branch at the War Office, the Military Aeronautics Directorate. He was therefore the first Chief of the Royal Flying Corps and surely, in all fairness, the real father of the Royal Air Force. Or perhaps, in view of Trenchard's later enormous contribution, we should acknowledge him in that rôle and revere Henderson as its grandfather.

For that year's manoeuvres, No. 3 Squadron's aircraft were made up to war strength of twelve. One crashed and was written off. Two were hauled back to Netheravon, unserviceable. Four forced-landed in enemy territory. Unfortuntely, the strategy and tactics of the exercise allowed scant opportunity for the use of the air, but valuable lessons were learned. One of these was that accurate observation did not have to be done from a few hundred feet, but was possible as high as 6000.

The squadron also practised aerial gunnery that year, to prepare for the time when aircraft would have to be armed and join combat. After much air-to-ground firing, and air-to-air with balloons as targets, the American-designed Lewis gun proved the most suitable machinegun: but, as usual, there had to be many months' wait for deliveries and none was received until after the outbreak of war.

No. 4 Squadron was also, by this time, at Netheravon, and presently No. 5 Squadron was at Dover.

No. 2 Squadron had meanwhile been posted to Montrose, on the east coast of Scotland, some twenty-five miles north-east of Dundee. The logic of this is obscure, for the area hardly enjoyed conspicuously good flying weather.

Its remoteness, however, gave rise to some pioneering long-distance flights. In August, Captain Longcroft, carrying Lieutenant Colonel Sykes in the passenger cockpit, flew a BE fitted with an auxiliary petrol tank from Farnborough to Montrose in seven hours forty minutes, with one refuelling stop.

The squadron took part in the Irish Command manoeuvres, which entailed flights of over 400 miles there and back. During the exercise a further thousand miles were flown without a single engine failure. The squadron's general efficiency was of the highest and it flew in all weathers. A touch of farce was seldom absent even from this all-weather aviating. Sometimes the wind speed was greater than that at which the aircraft could fly. The pilots then delighted in "tortoise races" in which the winner was the one who was blown farthest *backwards* over the course.

Major Burke, the squadron commander, had from the outset the sound order of priorities that has become established in every air force in the world: flying has to have precedence over all else. One of his diary entries reads: "... though barracks must be kept spotlessly clean, this work must be done by the minimum number of men, in order to swell the numbers of those available for technical work and instruction." It seems incongruous that, despite the clarity with which he saw what needed to be done and the efficacy of his leadership, he was an indifferent pilot who crashed frequently.

On all squadrons, instruction included map-reading, signalling, propeller-swinging, car-starting, and military and technical training. The pilots were engaged in practice reconnaissance, artillery co-operation, cross-countries and night flying. This last was in its infancy. Lieutenant Cholmondeley's flight by moonlight from Larkhill to the Central Flying School at Upavon on 16th April 1913, in a Farman, was the RFC's first.

The beginning of 1914, therefore, found the RFC efficient, possessed of an intelligent appreciation of the developments that must be expected in the event of war, and, although still small in numbers of pilots and aircraft, already vibrant with the inexhaustible vitality that was the hallmark which it handed on to the RAF.

The general restlessness in Europe heightened the feeling that war must be imminent and prompted the need for some special event to prepare the flying Service for it. Accordingly, June was devoted to an assembly of all the squadrons at Netheravon, for what was called a concentration camp. The mornings were spent in experimental work and trials; the afternoons at lectures and discussions. There were exercises in reconnaissance and in co-operation with other arms; in photography, balloon handling, and moving between landing grounds. Rivalry between the squadrons bred a strengthening of the comradeship and common loyalty that united them all in the one Service.

When, five weeks after the concentration camp dispersed, the RFC was mobilised, it was in all respects ready for war.

No. 2 Squadron's aircraft took off from Montrose for Farnborough on 3rd August, hours before war was declared, and eventually reached Dover by the 12th, after sundry accidents. The ground party left by rail on the 8th to embark near Glasgow for Boulogne. At 6.25 a.m. on 13th August, Lieutenant H. D. Harvey-Kelly was the first to take off from Dover and the first to land at Amiens one hour and fifty-five minutes later.

No. 3 Squadron's aircraft left Netheravon for Dover on the 12th, while the road party for Boulogne was embarking at Southampton. One pilot, with an air mechanic aboard, crashed and both were killed.

No. 4 Squadron had recently been transferred to Eastchurch. Its pilots also flew to Dover, and the remainder boarded ship at Southampton, on the 12th.

No. 5 Squadron was delayed by a dearth of shipping and by accidents to some of its aircraft. It flew from its new base at Gosport to Dover on the 14th and on to France the following day.

It was not until 22nd August that all aircraft caught up with their squadrons. There were unexpected hazards apart from those of mechanical failures and adverse weather. Lieutenant R. M. Vaughan, of No. 5 Squadron, was the last to arrive. He had taken off on the 15th but made a forced landing near Boulogne. Aircraft bore no national markings. Despite the pilot's uniform ("*Comment? La RFC? Ça c'est quoi, alors? Je n'en pige quedale. Je m'en fiche de votre Air Eff Say, Monsieur*") and other means of recognition, the French immediately arrested him and kept him in a police cell for a week: until he had satisfied them that he was not an enemy. Claiming to be an ally would cut no ice with them: they probably took it as a mortal insult to suggest that *La France* needed help from anyone in yet another conflict with *les sales Boches*. A poor advocacy for *l'entente cordiale*.

Nos. 6 and 7 Squadrons were in the process of being formed at Larkhill.

The responsibility for airships had been handed over entirely to the RNAS on 1st January, so No. 1 Squadron had parted with its two and existed, temporarily, on paper only: but was now immediately to be reactivated and equipped with aeroplanes.

Brigadier General Henderson and his Headquarters arrived at Amiens on 13th August. The Aircraft Park arrived there on the 21st. Twenty aeroplanes had been allocated to it, but of the nine BE2s, one BE2C, three BE8s (called "Bloaters"), and three Farmans, it had to supply half to the squadrons to bring them up to strength. It also took four crated Sopwith Tabloids.

The Royal Flying Corps, an amalgam of officers and men from diverse regiments and corps, bringing with them a great diversity of customs, traditions and uniforms, and equipped with a gallimaufry of six different types of aeroplane, had, none the less, begun to establish its own style and traditions. It was already a cohesive entity that was proud of itself and full of confidence.

CHAPTER 4

1914. At the Front

In the matter of leadership, with Henderson in command at the Western Front, the British were better placed than ally or enemy.

It was rare for a man who had been primarily responsible for the creation of a new arm to lead it in battle; and so soon after its inception. In Henderson, the RFC had a commander who was both intellectual and practical. He had studied Engineering at Glasgow University; but found his recreations in reading poetry and prose, writing songs, playing the piano, trying to compose music, inventing plots for short stories and working on a biography of the Black Prince.

After university, his father sent him around the world. He returned broad-minded, self-reliant and with a thirst for travel that never left him. He was to have ample opportunity to slake it. When, in 1883 at the age of twenty-one, he joined the Argyll and Sutherland Highlanders, the battalion was already preparing to set off for Cape Town. He immediately impressed his brother officers with his charm and his ability to take on any task and do it better than anyone else. In 1884 he saw action for the first time, against the Zulus. The next year the battalion went to Ceylon, and, three years later, on to Hong Kong. It returned home in 1892.

In 1898 Henderson was ADC to Brigadier General Lyttleton at the Battle of Omdurman. In 1900, during the South African War, he was wounded at Ladysmith when a brevet colonel on Kitchener's staff.

In 1904 the War Office published his *Field Intelligence*, which became an official textbook. He further increased his military reputation three years after that with the publication of *The Art of Reconnaissance*. In

1907 he became Staff Officer to the Inspector General of Forces; and accompanied Field Marshal Sir John French – who commanded the British Expeditionary Force in 1914 – to India and Malaya in 1909, then to Canada in 1910. When, therefore, he qualified as a pilot in 1911, he was a well-rounded, sophisticated and popular man of many parts, with all the social graces to complement his natural talents and considerable intelligence.

It must have been the artistic side of his nature and his love of books, as much as shrewdness and foresight, that prompted him to appoint as his personal assistant a man who seemed an unlikely candidate for such an appointment but proved surprisingly efficient in practical matters, and later as constant to Trenchard as to Henderson. This was the Hon. Maurice Baring, forty-year-old fourth son of Lord Revelstoke, who had forsaken the Diplomatic Service to become a foreign correspondent, war reporter, travel writer, novelist and critic. His only apparent credential for his duties at Henderson's right hand appears to have been competence in seven languages, including French, German, Russian and Italian. Although, when war was declared, he went at once to see Henderson, whom he believed to be still Director of Military Training, to volunteer for the Army and service at the Front as an interpreter, he did so only because they had known each other for seventeen years; not because his old friend was head of the RFC. "I did not know what the Flying Corps was or that there was a Flying Corps," he confides in his memoirs.

As an example of the most fatuous tautology, this avowal is hard to equal. If he were unaware of the Flying Corps's existence, it is self-evident that he could not possibly know its function. How a newspaperman, particularly one who reported wars, could be unaware of the creation of the RFC is not readily intelligible. Perhaps his protestation was a pose. Neither ignorance of basic facts nor affectation seems a desirable quality in someone aspiring to a position of military trust.

Henderson offered him little hope but promised to do what he could. Four days later, at six in the evening, Baring had a note from Henderson telling him to report to the War Office the following morning, prepared to accompany him to France. The casual abruptness of this summons, the instant transformation of a middle-aged aesthete of inveterate civilian habit into a military officer convincing enough to play the part of Commanding General's amanuensis, is breathtaking. But such strange reversals of rôle and revelations of hidden facets of character are not uncommon in wartime. Perhaps it is one of the qualities of a natural leader that he can discern the potential of an eager amateur and rate him "for the duration" equal to the professionals.

Early next day, a Sunday, Baring was appointed a lieutenant in the

Intelligence Corps and attached to RFC Headquarters. In the expectation of obtaining some sort of military employment, he had already bought "some khaki on which it was only necessary to put badges of rank in order to make it into a kind of uniform". This garb presumably consisted of a tunic and breeches; and, one supposes, a headdress. A pantomimic scene ensued when he donned these garments: "Six people endeavoured to put on my puttees. None were successful except finally in the evening, Sir David Henderson." Further evidence of the Brigadier's informality, this evokes a hilarious mental picture of the woebegone bedraggled pseudo-subaltern being dressed by his Commander-in-Chief while the baffled and apparently bungling captains and majors look on.

At nine that night Major Salmond called for Baring and they were driven to Farnborough, where they slept in a billiards room crowded with officers, at the Queen's Hotel. No time-wasting, he was pitched in at the deep end.

The morning brought a fresh confrontation with his puttees. He rose at five-thirty to wind them on; so tightly that he could hardly walk. Then to Farnborough station to take a train to Redhill, his companions, Salmond and Longcroft, talking mystifyingly about "bumps, pancakes, stalling and taxiing", exactly as airmen today would bewilder laymen with such technicalities as flame-outs, Derry turns, after-burners and bunts. Arrived at Newhaven, they found that they had to stay overnight before sailing for Boulogne: so, at Longcroft's suggestion, and in spite of Baring's trepidation about further struggles with his puttees, they "bathed in the dazzling sea". That night they slept in a railway carriage. Next morning there was discussion about whether the new officer was a lieutenant or second lieutenant. The former was decided upon and he duly went to a tailor on the pier to have his badges of rank sewn on.

They arrived at Amiens in time to spend a night in their bedding rolls on the grass of the aerodrome and to splash in their portable canvas baths next morning to the plaudits of spectators, before Lieutenant Harvey-Kelly and the rest touched down.

The French Air Service, meanwhile, had not only been reconnoitring but had also, on 14th August, bombed the German airship sheds at Metz. The Germans, in their turn, were already penetrating as deeply as 200 miles into France on reconnaissance.

The Army Council's logical decision that, because Henderson had nursed the RFC from birth, he should lead it in the field, was a great disappointment to Trenchard, who, reasonably, expected the Head of the Service to be retained at the War Office while someone of his own rank, experience and vigour was sent to the battlefield.

Like Henderson, of whom he was the antithesis in most ways, he was a reader as well as a doer. His preference was for biographies of men of action: Clive, Warren Hastings, William Pitt, the Duke of Wellington.

He was, however, given a task for which he was no less well suited than for one at the Front. Henderson put him in command of the Military Wing – still an official title – at Farnborough, charged with the prime duty of forming the new squadron needed for the RFC's rapid expansion.

Lieutenant Colonel F. H. Sykes, who had been a conspicuous figure in military aeronautics from the earliest days, loathed Trenchard, and was commanding the Military Wing when war broke out, had become Henderson's Chief of Staff. Cold, unpopular, an intriguer, he was the object of general mistrust but had contrived to win Henderson's confidence.

Sykes, visiting Farnborough, made the mistake of officiously telling Trenchard that his main responsibility would be to train replacement pilots for the squadrons in the field. This ignited Trenchard's short fuse and exploded a typical rasping reproof. "Don't talk such damned rubbish. My job here, as you should know, is to produce new squadrons."

With such a forthright, uncompromising and incisive man to support him, Henderson was able to depart for France, after seeing his squadrons off from Dover, confident that he had left all well at home in the best possible hands.

There were many contradictions in Trenchard's nature. He was a genius at administration and organisation, although academically a dolt. From his preparatory school he went straight to a crammer's at the age of eleven, to prepare for the entrance examination for the Royal Naval College, Dartmouth. He failed and was sent to another crammer to work for the RMA, Woolwich. He failed that examination twice. He then sat thrice for the Militia, a back-door means of entry to the regular Army. The pass mark was 1645 out of 2400. At his third attempt, in March 1893, he scraped in at last with 1673.

Two months later he sailed for India as a platoon commander in the Royal Scots Fusiliers. At once he became conspicuous for his strength of character, dedication to his regiment and platoon, and determination to excel at all that he undertook; and for his bumptiousness. If Trenchard wanted to pursue some activity for which no facilities existed, he created them. Having asked his Colonel for permission to form a polo team, within six weeks he had persuaded other officers to buy horses, had established stables and obtained a playing ground. He applied himself to shooting so assiduously that in 1894 he won the All-India Rifle

Championship gold medal and the Viceroy's Cup. More importantly, he coached his platoon until every man was a first-class shot. When he took three months' leave he organised race meetings, parties and balls wherever he went. He was a restless man of boundless energy, who had the delicacy of a steamroller when faced with opposition or reluctance. A born high-powered busybody.

He was also fearlessly pugnacious in defence of himself. Subjected to constant bullying by a major, he waited until the battalion went into camp: then loosened the guy ropes of the major's tent one night; and when the infuriated senior officer emerged from the enveloping folds of tumbled canvas, threw a jug of cold water in his face for good measure. This cost him a reprimand. It was a brave deed that could have incurred a court martial.

In turn, he was not slow to behave badly himself. As a major, he once greeted a newly arrived subaltern by asking him if he liked riding. The subaltern said he didn't. "*Good!*" Trenchard exclaimed, and ordered the newcomer to meet him at the stables after tea: taking advantage of a junior from whom military discipline demanded obedience. His glee at the prospect of making someone suffer, which is implicit in "good", is unendearing.

Boorishness figured among his less attractive qualities. His hostess at a dinner party mentioned that Trenchard was her maiden name and produced a family tree for his diversion. "There are only two branches of the family that interest us," he told her. "The main one, to which I belong; and a second one which I have always understood was founded about two hundred years ago by the illegitimate son of an umbrella manufacturer in Manchester." The party thereupon broke up, but Trenchard was too pleased with himself to show remorse. His host wrote to his Colonel, who duly reprimanded him again for his atrocious manners. There was also a distastefully snobbish intent behind the reference to the umbrella-maker.

He was insolent. In India, after the murder of a British Commanding Officer, Lord Curzon, the Viceroy, ordered that he must be sent reports of all military occurrences, however minor. Trenchard, when Orderly Officer, sent Curzon a telegram to say that he had rebuked a sergeant for throwing a lump of fat at a punkha coolie. (A servant who pulled a rope to swing one of the cloth fans that hung in every room.) In South Africa, when a general ordered pickets to be posted on some hills, and a major informed Trenchard that the general was interested in every detail of this, Trenchard, exercising his brand of humour, asked whether the general wanted fat or thin men in certain positions.

If Sykes had unattractive qualities and Trenchard could be as unprepossessing, Henderson in contrast was charming, as well as straight, honourable, scrupulous, loyal and clever. Trenchard shared the first four of these virtues and was clever also, but in a different way. He was, however, like many military leaders of the time, inarticulate in both the written and spoken word.

These two largely disparate men respected each other and worked in harmony. Their immediate target was to increase the number of squadrons to twelve. A less sanguine estimate of the numbers needed successfully to oppose the enemy was made by the officer who, as Deputy Director of Military Aeronautics, was in charge of supply and equipment at the Military Aeronautics Directorate, Lieutenant Colonel W. Sefton Brancker. He thought the target should be thirty squadrons.

Sefton Brancker had been hovering in the background of military aviation for over three years, had moved into the foreground and was worth paying attention to.

In style he was the quintessential cavalryman, a monocled dandy whose quiet charm and self-deprecating manner camouflaged a dashing spirit. He described himself as "a very moderate pilot", but, in June 1914, flew a BE2c, the first aeroplane with true inherent stability, from Farnborough to Upavon and wrote a report while he did so: flying hands-off, except when he had to use the throttle. He first appeared on the scene in January 1911, when, stationed in India, he chartered an aeroplane to take part in the cavalry manoeuvres. Flown by a Monsieur Jullerot, he, as observer, did some useful reconnaissance: until the second day, when the aircraft crashed and fell to pieces. In 1913 he learned to fly, attended CFS, was put on the RFC Reserve, and presently replaced Captain Ellington at the Military Aeronautics Directorate. He now began to play an increasingly important part in the Service.

And it was not long before no less a potentate than Kitchener doubled his estimate of necessary squadron strength to sixty.

For the time being, there were only 4 squadrons available, comprising 63 aircraft. The French started the war with 23 escadrilles, each of 6 aeroplanes: a total of 138. The Germans sent 198 aeroplanes to the Front, in 33 Field Service units, each 6-strong, and retained 10 Home Defence units, of 4 aeroplanes each. The performance of the three countries' aeroplanes was about the same, and only the French Voisin III bomber had a permanently mounted machinegun.

The fighting spirit of the British pilots would have to compensate for lack of numbers. Before No. 2 Squadron left Dover, Major Burke (nicknamed "Pregnant Percy") told his pilots that, although they were unarmed, he did not expect them to ignore any Zeppelins they might happen to see en route: a subtle but clear order to ram. From the first day, the feeling of being engaged on a hazardous adventure was cognate with recognition that this was a matter of highly practical ruthless killing and self-sacrifice. All pilots and observers wore revolvers. Many took rifles up with them. Before leaving England, some cockpits had been fitted with racks to hold a rifle and a few grenades. In addition to the twenty-pound Hale bombs that were stowed in the cockpits and dropped over the side by hand, fléchettes – steel darts five inches long and three-eighths of an inch in diameter – were carried, 250 at a time, in canisters under the fuselage from which they were released by tugging a wire. They, grenades and bombs were all used against other aircraft as well as ground targets. These crude attacks were not so much hit or miss as miss or miss. The chances of hitting a stationary target from an aeroplane travelling at sixty miles an hour were remote; of hitting another aircraft, when its slipstream as well as its speed would complicate judgment, they were about equal to those of the biblical camel's of passing through the eye of a needle.

At first, however, both sides were mainly engaged in reconnaissance. The British were not ready to start at once. Amiens was only a staging post on their way to what was intended to be a permanent base. On the 16th, Headquarters and Nos 2, 3 and 4 Squadrons moved to Maubeuge. But they did not have to wait until they began operational flying to suffer their first losses. Lieutenant E. W. C. Perry and his mechanic crashed in a burning BE8 and were killed. Two days later No. 5 Squadron joined the others and Second Lieutenant R. R. Smith Barry broke several bones when his BE8 crashed. His passenger, a corporal, died. An unexpected menace came from their own side. Flying over French troops was an invitation to be fired upon with rifles. After a few days the British Expeditionary Force appeared in the area, and now it became equally risky to fly over one's own people.

The atmosphere seems to have been light-hearted and unresentful, since nobody was actually hit. Lieutenant R. S. Wortley, reminiscing, rhapsodised: "Pilots – one couldn't find a jollier lot of fellows. Such exuberance of spirits I have never seen. The cool confident courage with which they handle these very imperfect machines of theirs and the zest with which they respond to any and every call upon their services is quite amazing. Hardly a day passes but our machines are fired at from the ground: not only by French sentries but by our own infantry as well."

The RFC flew its first operational sorties on 19th August, when, at 9.30 a.m., Captain P. B. Joubert de la Ferté, a flight commander on No. 3 Squadron, flying a Blériot, and Lieutenant G. W. Mapplebeck of No.

[34]

4, in a BE2, took off on the RFC's first reconnaissance. Joubert de la Ferté's task was to find out if Belgian forces were in the Nivelles–Genappe area. Mapplebeck's was to see if enemy cavalry were in strength near Gembloux. They were supposed to keep company as far as Nivelles, so that if one were to make a forced landing the other could report where he had done so. The weather was cloudy and deteriorating. The pilots lost sight of each other and soon neither knew where he was. Mapplebeck arrived over Brussels, but did not identify it! Eventually he did find Gembloux but saw only a small force of German lancers. After groping his way through cloud and following the Sambre, he landed back at Maubeuge at midday. Joubert de la Ferté, in a slower aeroplane, strayed about between cloud layers until he landed at Tournai, where he could learn nothing about the Belgian Army, so flew on until, lost once more, he put down at Courtrai. The Gendarmes told him that the Belgian Flying Corps Headquarters was at Louvain. Although this did not entirely fulfil his brief, it was the best he could do in weather that was scarcely fit to fly in. He returned to base at 5.30 p.m. The sorties were not without value and Henderson telephoned their results to Field Marshal French.

Baring recorded that "The weather was fine and hot. There was no excitement. Life was like a cheerful picnic."*

It was not quite a picnic for everyone, however. On the 22nd twelve reconnaissances were flown, in the face of machinegun as well as rifle fire from the ground, and yielded much valuable information: the Germans were advancing, leaving burning villages in their wake. The day was notable for three occurrences, each the first of its kind on record. Sergeant Major W. S. Jillings, of Two Squadron, observing for Lieutenant N. Noel, was shot in the leg: the first British airman to be wounded in flight. The day's major excitement was provided by a Taube, the type that constituted half the German aircraft strength. It was spotted approaching the aerodrome at 4000 feet and Lieutenant Louis Strange, of Five Squadron, who had been allowed to fit a Lewis gun to a Henry Farman F20, took off to shoot it down. The gun was so heavy that he was unable to overtake it or climb beyond 2000 feet. This was the RFC's first "scramble" and the armed Farman was the earliest ancestor of modern interceptor fighters. The third "first" was an unhappy one: another No. 5 Squadron aircraft, piloted by Lieutenant V. Waterfall, with Lieutenant C. G. G. Bayly as observer, was brought down over enemy territory by ground fire and its crew taken prisoner.

The day's last sortie brought back a report that sent Henderson

* The "Phoney War", the "Sitzkrieg" experienced by the BEF and the RAF squadrons in France in 1939, was much the same.

[35]

hurrying to General Headquarters to deliver it personally to Field Marshal Sir John French. All day, reconnaissance had found enemy forces on the move in what appeared to be an attempt to surround the Allies. Now came confirmation that a whole Army Corps was travelling west along the Brussels–Ninove road; and, aware already of the efficacy of aerial reconnaissance, hiding as much as possible under the trees that lined both sides.

The Germans planned to sweep into France through Belgium and Luxembourg: but, instead of concentrating their weight on their right wing, they gave equal strength to their left so as to meet an expected French offensive in Lorraine. The result was that the French met stronger resistance than expected and were forced to fall back: while the Germans' right flank was held up by a much stronger Belgian defence than it had reckoned on. The British were also facing the enemy's right, near Mons.

Field Marshal Joffre ordered a retreat along his whole front. In consequence, Field Marshal French decided not to mount his intended offensive but to hold the line at Mons, which was soon under attack.

On the 24th a general Allied retreat began. RFC HQ and its squadrons moved to Le Cateau. In the next ten days they moved eight more times: maintaining, it seems, a light-heartedness that belied any suggestion of calamity: although there was some chaos, caused by crowded roads and mistakes in drivers' map-reading. Officers and men ate whenever, and slept wherever, they could. The first night, the pilots and staff officers spent in a straw-filled barn. Joubert de la Ferté remembered that one night he slept under a hedge in the rain, the next comfortably in a private house, the following one in a luxurious hotel. Louis Strange recalled: "The usual orders on the retreat were dawn reconnaissances, dropping hand grenades and petrol bombs on the enemy, and when it was impossible to notify pilots of the next aerodrome, the orders were to fly approximately twenty miles south and look out for the remainder of the machines on the ground, if machines had left the last aerodromes."[*]

McCudden recorded that at six o'clock on the morning of the 26th officers and men breakfasted together around the field kitchen; and, on another day, Lieutenant Conran, one of the pilots, milked a cow in a roadside field when the squadron's convoy halted for a midday meal. He also mentioned working all night as a matter of course, servicing the aeroplanes: as ground crews have done ever since.

Joubert de la Ferté had something else to write about the turmoil of those ten days that was also repeated in May and June 1940. "What we

[*] The haste and hurly-burly, the confusion and uncertainty were all to be experienced again by the RAF squadrons and Headquarters in their retreat across France in May and June 1940.

saw during the advance confirmed our impressions from the air as to the unspeakableness of the Hun in his methods of dealing with the civilian population. I saw half a dozen villages on fire during the first day of the battle, twenty miles west of Mons, where by no possible means could there have been any armed resistance to the passage of the Huns. It was simply frightfulness on the part of the Uhlans, and what we saw later was a clear indication of wilful and unnecessary destruction of private property."

German airmen meanwhile bombed dwellings and refugees. Such evidence should be remembered when reading tributes to the chivalry of German pilots, who are often credited with good sportsmanship and an admirable sense of decency. They belonged to the nation that deliberately provoked war with France in 1870, determinedly set about stirring up the war that it declared in 1914, and spent years plotting the war it brought about in 1939.

The airmen who perpetrated these enormities were the same who used to drop wreaths on Allied aerodromes in maudlin tribute to the airmen they had killed, and messages reporting those who had been taken prisoner. Mawkish sentimentality is never edifying. Transports of bogus contrition are an insult to the intelligence.

On the 25th, Harvey-Kelly, of Two Squadron, scored the first victory in aerial combat, without firing a shot. Having intercepted a Taube, he was presently joined by two more BE2s. He positioned himself a few feet astern, followed it through all the desperate weaving and switch-backing with which its pilot tried to evade him, and forced it down near a wood: into which the German pilot and observer bolted. Harvey-Kelly and Lieutenant W. H. C. Mansfield landed near the enemy aircraft. They searched unsuccessfully for the fugitives, then burned it.

In the course of the almost daily stumbling south-westward under the enemy's onslaught, the RFC found itself on 31st August at Juilly. Baring started this day by buying "a beautiful resounding brass bell" at Senlis, for the mess. In the afternoon he set off by car with "Colonel" Burke, as he calls him, who was actually a major at the time, and "Captain" Crosbie, who was in fact then a lieutenant. He put up yet another of his inane performances. "I was given a map and told to direct the proceedings. I didn't know how to read a map." After taking the wrong direction at a crossroads, followed by hours of quartering the countryside and asking the way, the three officers eventually found their destination. On the morrow they enjoyed "a peaceful day at Juilly and a nice bathe in a great pond ..." but the picnic atmosphere was spoiled at nine-thirty that evening, when panic broke out. Headquarters, two miles away, thought they had been cut off by the Germans, which meant hasty packing and departure.

Henderson's unruffled deportment brought an admiring tribute from Seeley. "One evening, returning from Smith-Dorrien's front with a despatch to Headquarters, at the worst moment of the retreat, when the German advance guard was within thirty-five miles of Paris, and passing the temporary aerodrome which was the Headquarters of the Flying Corps, for the moment, I saw David Henderson standing on the bank, looking down on the road surveying the rapidly moving motor-cars, lorries and troops and said to him 'When are you moving on?'

"He had the same unperturbed demeanour, the same kindly and whimsical smile, which we all knew so well, and he replied, 'Well, we have a few more aeroplanes to come in and we shall fly away at dawn.'

"I said, 'But suppose the advance guard gets here before dawn?'

"He replied, 'Then we shall get away in the dark as Graham-White did when he was flying from London to Manchester!' [In 1910 the *Daily Mail* offered a prize of £10,000 for the first flight from London to Manchester. Claude Graham-White attempted it in a Farman but had to forced-land near Lichfield, with engine trouble; where high wind in the night damaged his machine.]

"He was not in the slightest degree perturbed or flustered. I have never seen a man at a desperate moment so completely master of himself. His cool courage always communicated itself to all who were with him."

Five days later the Battle of the Marne began. French ordered a significant repositioning of the RFC that would soon become standard. Reconnaissance having been recognised as vital, aeroplanes were to operate directly with First and Second Corps. The CO of No. 5 Squadron was detached, with three aircraft, to Sir Douglas Haig, No. 3 Squadron's OC and three aircraft went to Sir Horace Smith-Dorrien. A wireless-equipped aircraft from No. 4 Squadron went to keep RFC HQ supplied with information. Decentralisation of the RFC had begun.

The Imperial German Military Aviation Service had been thus organised from the outset, with an aeroplane section at First Army HQ for long-range strategic reconnaissance, and sections for tactical reconnaissance attached to the various corps.

The pell-mell retreat and the nearness of the enemy to Paris did not cast any great gloom on the British airmen, who were in high spirits, full of energy and horseplay. Some of the squadrons were billeted in a girls' school, where Baring slept on the first night of the Battle of the Marne. He described a pillow fight in which the pilots dressed themselves up in the girls' nightgowns. One of the dormitories was invaded and fell only after a vigorous defence. That sounds like a display of the ebullient

[38]

spirit which has animated countless officers' mess parties down the decades.

On that day Henderson wrote to his wife: "My people are so good that I get a lump in my throat when I think of all they have done. We have lost pretty heavily. These Germans are certainly determined fellows. The soldiers are nothing out of the way, only there are a desperate lot of them, but their generals are most resolute fellows with any amount of push and moral courage. But they take heavy risks and I hope that one of these days we shall catch them on the hop."

The day was only seventy-two hours off. By evening on the 9th September the enemy was in retreat all along the line and the battle had been won.

On the 10th Henderson wrote home again: "I have just had a note from HQ giving me a message from General Joffre, the French Commander-in-Chief, to our Field Marshal, thanking him for the wonderful information supplied by the English Flying Corps. It is to be published in Orders tonight, I believe. I understand that nobody at home knows anything about our work, but it will all come out soon enough. Maurice [Baring] is a treasure. He goes poking about among the French people ..." Modern idiom makes the mind reel at this libidinous implication. But one is quickly reassured. The General imputes no Priapic antics among the French ladies to "Maurice". "... to find a café where I can have a good dinner and is always smiling and cheerful.

"The casualty lists must be pretty bad reading. We shall lose a lot more before this war is over, but you can't make an omelette without breaking eggs." Such platitudes from a well-read man who is a published author are unexpected. There is a surprising hint of callousness too: perhaps no more than a symptom of the resignation that all generals, admirals and air marshals must acquire to the necessity of sending men to their death; and acceptance that the patriotic end justifies the means.

Heavy loss of life did not affect his appetite. "I have had to draw £15 to keep me going. I try to feed well so as to stay healthy, but I fear that if the advance continues we won't get much to eat except our rations."

His son had recently gone from Eton to the Royal Military College. "I do hope the little man will do well at Sandhurst. I wish I could go and see him there. I have always looked forward to inspecting my cadet son." Sentiment comes uncomfortably close to sentimentality in his first sentence.

Joffre's Order read: "Please express most particularly to Marshal French my thanks for the service rendered to us every day by the English Flying Corps ..." Henderson might have winced at that; along with the other Scots, Welsh and Irish in the RFC. "... The precision, exactitude and regularity of the news brought by them are evidence of their perfect organisation and also of the training of pilots and observers."

French, in his first despatch, wrote: "I wish particularly to bring to your Lordships' attention the admirable work done by the Royal Flying Corps under Sir David Henderson. Their skill, energy, and perseverance have been beyond all praise. They have furnished me with the most complete and accurate information, which has been of incalculable value in the conduct of operations. Fired at constantly both by friend and foe, and not hesitating to fly in every kind of weather, they have remained undaunted throughout. Further, by actually fighting in the air, they have succeeded in destroying five of the enemy's machines."

The Allies pressed forward across the Aisne and heavy fighting continued daily. Artillery observation became of equal importance to reconnaissance. The RFC reported where enemy batteries were, then controlled the counter-fire of their own gunners. Two officers of Three Squadron were outstandingly effective at this work. A wireless tele-graphy unit commanded by Major H. Musgrave had been formed. Lieutenants D. S. Lewis and B. T. James flew alone in BE2s with a wireless set to report and correct the fall of shells, usually under heavy fire themselves.

On 22nd September, Mapplebeck, bombing an anti-aircraft battery, met rifle fire from an Albatros. Wounded in the leg, he became the first British airman to be hit from an enemy aircraft.

On 27th September Smith-Dorrien, in a signal to French, expressed "... my great admiration for the splendid work the Royal Flying Corps is doing for my Corps from day to day. Nothing prevents them from obtaining the required information, and they frequently return with rifle or shrapnel bullets through their aeroplane or even their clothing, without considering such, to them, ordinary incidents as worth mention-ing. Today I watched for a long time an aeroplane observing for the six-inch howitzers for the Third Division. It was, at times, smothered with hostile anti-aircraft guns, but nothing daunted, it continued for hours through a wireless installation to observe the fire and indeed to control the Battery with most satisfactory results. I am not mentioning names, as to do so, where all are daily showing such heroic and efficient work, would be invidious."

It had needed a mere six weeks at the Front for the RFC to overcome the prejudice which, only a couple of years earlier, had dismissed it as useless; and alarming to horses.

Anti-aircraft guns were proliferating and had already earned a facetious nickname, "Archie", from a popular song in which the phrase "Archibald, certainly not!" repeatedly occurred. It originated when

Lieutenants A. E. "Biffy" Borton and F. G. Small, his observer, of No. 5 Squadron, came under heavy fire. Borton made a succession of 45-degree turns to dodge the shell bursts, and they derisively shouted the catch phrase each time.

Photographic reconnaissance began on 15th September, when Lieutenant G. F. Pretyman of No. 3 Squadron took five pictures of the enemy line. These were the first of the thousands that would eventually form a map of the whole Front.

"This September was a beautiful still golden month," wrote Baring, "and I used to spend a great deal of time with the squadrons."

Later: "I remember the heat of the stubble on the Saponay aerodrome: pilots lying about on the straw; some just back from a reconnaissance, some just starting, some asleep, some talking of what they would do after the war."*

Less idyllically: "A tragic incident occurred ... One of the pilots was practising signalling and dropping lights [firing Vérey lights]. He was flying quite low over our trenches backwards and forwards." The machine, as so often occurred later during the war, was thought to be behaving in a "suspicious manner" and was fired at by our troops, and before they could be stopped it was brought down amid their cheers. "When the machine crashed they saw that the pilot was an Englishman and that he was dead."

A phase of the war was over. Constant movement and fluidity had given place to static warfare. The adversaries were digging in. They faced each other across no-man's-land in trenches, separated by distances from a hundred yards to a mile, which would, by the end of the year, stretch from the Belgian coast to the Franco-Swiss frontier. The French and Germans were already using kite balloons to supplement their aeroplanes for observation of enemy movement and for artillery control. The scope of aerial activity by both sides was broadening.

During these weeks, when British or French aircraft passed a German machine, the occupants waved at each other. Sometimes a pilot would show off with a vertical bank, a steep dive and zoom (which the RFC called a "hoick"), by rocking his wings, or some other display of skill and the aircraft's manoeuvrability. "Stunts", such as looping, were attempted only by a few specialists who performed at air shows. Nobody yet knew how to recover from a spin. Rolls were shunned. The enemy would respond with some mild act of aerial flamboyance before both continued about their business with no ill feeling. Even if they exchanged

*He might have written the same about a fighter airfield in the early 1940s, substituting only "grass" for "straw".

[41]

shots with revolver, pistol, rifle or shotgun, it was more as a gesture than in any expectation of inflicting damage to aeroplane or occupants.

These encounters became less genial after the 5th October when an Aviation Militaire mechanic, Louis Quénault, behind a Hotchkiss machinegun aboard a Voisin flown by Sergeant Joseph Frantz, shot down an Aviatik: another feat that was the first of its kind, and the most important so far.

Henderson, whose quick mind had recognised the importance of the aeroplane as a military weapon four years before the war, telegraphed to the War Office a month before Frantz's and Quénault's revolutionary success: "There are no aeroplanes with the RFC really suitable for carrying machineguns; grenades and bombs are therefore at present most suitable. If suitable aeroplanes are available, machineguns are better undoubtedly. Request you to endeavour to supply efficient fighting machines as soon as possible."

After the Aviatik was shot down, he told Seeley: "This is the beginning of a fight which will ultimately end in great battles in the air, in which hundreds, and possibly thousands, of men may be engaged at heights varying from 10,000 to 20,000 feet."

The campaign in Flanders wore on towards the First Battle of Ypres. Bombing developed faster than combat between aircraft. The most efficient of the Royal Naval Air Service squadrons was sent to Dunkirk with three BE2s, two Sopwiths, two Blériots, a Farman, a Bristol and a Short seaplane whose floats had been replaced by wheels. On 22nd September the squadron bombed airship sheds as far away as Cologne and Düsseldorf. Some aircraft were detached to Antwerp. From there, on 8th October, strategic bombing was initiated when the same target at Düsseldorf was attacked again and completely destroyed, the flames rising to five hundred feet; and Cologne was also attacked for the second time, but mist hid the airship sheds, so bombs were dropped on the railway station.

German bombing was less effective. On 1st September a Taube had dropped bombs on Montmartre and killed four civilians. On 11th October, two bombs hit Notre Dame, and on the 12th, the Gare du Nord.

The Germans held the Allied counterattack against their assault at the First Battle of Ypres. RFC HQ and the squadrons were at St Omer. Bad weather with low cloud hindered reconnaissance. Inexperienced replacement pilots and observers sometimes misinterpreted what they saw. There was still no official category of "Observer". The man in the second cockpit was sometimes a pilot, sometimes a member of the ground crew, of whom, in No. 3 Squadron, McCudden was often one. No. 6 Squadron had been sent hastily to France. One day one of its crews mistook the darkness of long stretches of tar on a macadamised road for an enemy column. On another, when someone saw orderly ranks of shadows in what looked like a field, he thought he had found an enemy force in bivouac. But they were gravestones in a cemetery, not tents, that were set in rows.

Casualties on the ground were heavy, and Henderson was not too preoccupied by his concern for his squadrons to be affected by them. At luncheon one day he recounted how three Germans with a machine-gun, in a small house, had fought off repeated attacks all morning and killed a great number of British infantry. Two of them were killed and the third went on fighting, terribly wounded, until at last he killed himself.

Someone commented, "He deserved to live."

"He deserved to die," Henderson corrected him.

During the battle, French asked the War Office for more squadrons. These were promised as soon as possible and a proposal was made to reorganise the RFC into wings, commanded by lieutenant colonels, comprising three or four squadrons. A wing would be attached to each Corps. This would relieve the Commander of the RFC of much of his responsibility, as he would no longer be concerned with the field work of his squadrons. Instead of a distinct Flying HQ commanding, directing and maintaining the RFC, there would merely be a senior RFC officer on the Commander-in-Chief's staff, responsible only for maintenance of the Service: in fact the status of the RFC would be considerably reduced.

Henderson's reply agreed with decentralisation into wings, but wholly opposed abolition of the Flying Corps Headquarters. "In dealing with this highly specialised branch of the Service it is very desirable that the part of the Commander of the RFC in the field should be retained, and that this Commander should be in very close touch with General Headquarters. Without interfering in any way with the power of the Army Commanders to deal with the wings attached to them in such a manner as they think best, it is necessary that there should be uniformity of method and of operation throughout the Corps. The Commander therefore should have the power to issue instructions on all technical matters, including the technical handling of aeroplanes either for reconnaissance or fighting purposes, and he should also have full authority to change, move or replace officers and men of the squadrons as may be found necessary for the efficiency of the Corps."

French approved this and forwarded it to the War Office on 1st

November. The Army Council responded on the 18th that the principles were acceptable "but it appears to question whether, once intended decentralisation is completely effected, and is working satisfactorily, there will be a necessity for the permanent retention in the permanent organisation of General Officer Commanding RFC as proposed by you".

Henderson had been promoted to major general on 26th October. The reorganisation of the RFC that had been outlined would have no place for an officer of his seniority and achievement. A few days later he visited Farnborough. He found the volatile and short-tempered Trenchard disaffected by what he had heard from querulous officers who, returning from the Front, had complained that Henderson, over-influenced by the excessively cautious Sykes, had not made bold enough use of the squadrons so as to deny the enemy freedom of the air. In conclusion, heated as he all too often was, Trenchard said that he wanted to return to his regiment.

Fortunately for him he was dealing with a wiser and far more equable man, who overlooked the offensive implied criticism of himself and calmly explained the proposals of which Lord Kitchener, the War Minister, and Field Marshal French had approved. He told Trenchard that he wanted him to command the First Wing.

Instantly attracted, Trenchard would not accept until he knew what would be the egregious Sykes's place in the new organisation. Henderson answered that Trenchard would be junior only to himself, and Sykes would have no operational authority.

On 18th November Trenchard landed at Boulogne to take up his new command. Baring was sent to meet him; and once again made a booby of himself when they set off by car for St Omer, by letting the driver take a road that went in the direction opposite from their destination, and would have led them to the enemy lines, if a French officer at a road block had not turned them back. Their conversation cannot have instilled the new Wing Commander with confidence in him, either. "Colonel Trenchard asked me a great many pertinent questions, few of which I could answer." They arrived at 8 p.m. and the colonel had to sleep on the floor of the guest room. One wonders what aberration from his usual behaviour restrained Trenchard from usurping Baring's bed and consigning him to the hard lying.

Four days later the precarious balance of personal relationships was rocked by the shock waves of a tragedy at the Front. On the day that Henderson's promotion was gazetted, a shell mortally wounded Major General Lomax, who commanded the First Division. His second-in-command and temporary replacement did not satisfy the First Corps Commander, Sir Douglas Haig, who consulted French, who recommended Henderson. Henderson was appointed to command First Division and Sykes to take his place at the head of the RFC.

Trenchard knew that Henderson had not dishonoured his word, but the reversal of what he had been promised naturally sent him speeding to French's HQ, where he refused to serve under a man who was junior in years and seniority, and asked to be sent back to his regiment. Kitchener, when informed of these changes, turned down Sykes's aggrandisement, for lack of experience, and forbade Henderson's posting. "I want him in command of the RFC," he said to Brancker.

"Perhaps the RFC is not a big enough command for a general of his seniority," Brancker suggested.

"He thinks too small, does he? Then you tell him I am going to have a lot of Generals in the Flying Corps before I finish."

But an unhappy Henderson had to take up his new appointment and Baring chose to go with him; no doubt thoroughly disoriented and asking route directions every few miles. Sykes, true to form, was sure that Trenchard had been machinating against him and said so.

On the day of Henderson's move, 22nd November, Strange and his observer, Lieutenant F. G. Small, witnessed a bizarre incident that provided an amusing insight into the relationship between commissioned and uncommissioned ranks in the German Services. The two No. 5 Squadron officers, in an Avro, on whose upper wing, in defiance of orders, they had lashed a Lewis gun, were at 7000 feet when they spotted an Albatros 500 feet below. They dived at it from abeam, then flew ahead of it and slightly below, so that Small could fire back at it. After the enemy had received the 94 rounds from the two drums of ammunition, at 550 rounds a minute, the pilot lost his nerve and landed behind the British lines. Strange and Small, circling low over it, saw that it was riddled with holes, but the crew were both unhurt. In the German Air Service the observer was always an officer and in command of the aircraft. The pilot was regarded as a mere minion and was often a sergeant, corporal or even a private. On this occasion, as soon as the Albatros touched down, the observer leaped out, dragged his pilot from the cockpit, knocked him to the ground and set about kicking him vigorously until he scrambled up and tried to take evasive action; not, of course, daring to defend himself against his superior. He must have been thankful when British troops turned up and put an end to his trouncing.

French, resenting Kitchener's interference, procrastinated as long as he could about complying with the War Minister's edict that Henderson must not leave the RFC. But on 20th December, Henderson returned to command at RFC HQ and Sykes reverted to his former position.

Baring, for all his irritating ineptitudes and the wet impression he conveyed of himself in his early wartime diary entries, endears himself by his steady application to learning a job that was utterly in contrast with any he had done as a civilian, and for his loyalty to Henderson, and, later, to Trenchard. He writes touchingly of his and Henderson's brief exile. "I used frequently to walk or ride into Bailleul to see the squadrons which were there, and although I was very happy with the First Division, I suffered the whole time from RFC sickness."

This large, shambling, kindly, cultured and phlegmatic man could describe the war with originality, too. "German warfare is like Wagner's music. The German use every possible accessory . . . spies . . . Zeppelins, flame-throwers . . . smoke screens . . . just as Wagner uses every accessory . . . scenery . . . lights . . . over and above the music to heighten the effect of the music."

By the time that Henderson and his Intelligence officer returned to St Omer, First Wing, consisting of Nos 2 and 3 Squadrons, under Trenchard, and Two Wing, Nos 5 and 6 Squadrons, under C. J. Burke, who was now a lieutenant colonel, had settled down at their new aerodromes. Air operations had begun to follow the pattern, tactically and in the enlargement of their scope, that they would pursue and develop for the next four years.

CHAPTER 5

———◆———

1915. Warming Up

The BEF's two corps had been elevated to armies. The two Wings remained where they were, under the control of their respective Army HQs. Based at St Omer were No. 4 Squadron, No. 9 Squadron – formerly the Wireless Unit – commanded by Captain H. C. T. "Stuffy" Dowding,* and the RFC Headquarters Squadron.

The German Military Air Service, whose Abteilungen had been attached to the various Armies and Corps from the outset, was also still concentrating on tactics rather than strategy.

Air operations by both sides fell into three categories: reconnaissance, whose purpose was to seek signs of fresh activity, such as troop movements by road and rail, the building of supply dumps and fortifications, the disposition of artillery units; observation, by means of which, with the aid of photography, the progress of such activity was followed; contact patrols, to keep Headquarters in touch with the ground forces during a battle, when, in an attack, ground changed hands and the situation became confused: usually with salients thrusting through the opposing line, so that friendly and enemy-held territory became indeterminate except from the air.

Apart from the similar capabilities of their aeroplanes, the broad framework of their organisations and their functions, the three air forces had human attitudes and characteristics in common. The basic and most important common factor was that all aircrew were volunteers. This meant that, whether British, Commonwealth, French or German, they

* Later Air Chief Marshal Lord Dowding, GCB, GCVO, CMG, who, as Air Marshal Sir Hugh, commanded Fighter Command in the Battle of Britain.

were animated by the same brand of enthusiasm, they had all responded to the urge to journey in what Elizabeth Barrett Browning called "the tingling desert of the sky". As the air forces expanded, the cheerful enthusiasts of the kind who had created the heady atmosphere of Brooklands, Hendon and Reims, and the more stolid ambience of the German flying centres, hurried to join. They were bolder than the average. Much remained unknown about flying. Aeroplanes in flight offered many dangers. Engine failure could cause a fatal stall. A manoeuvre that over-strained the structure could break a wing. Parachutes existed, but were forbidden to military airmen.

Hence the amiable, although not amicable, exchanges of greeting between enemies who flew past one another, each on his way to spy on what was happening behind the other's lines. Both parties understood the other's difficulties in handling the technical demands of flight; and to each the other was a rival in skill as much as an enemy. On both sides, also, airmen were to some extent frustrated by the ignorance and prejudice of senior officers who had never been airborne, but were in authority over them: a burden and a grievance that gave them mutual empathy, whatever uniform they wore.

Field Marshal Lord Kitchener provided an instance of the rigidity of the martial mind which any pilot, whatever his nationality, would recognise. Before a visit to Farnborough, he had told Trenchard that he wanted to see aeroplanes fly past in formation. Trenchard explained that this was impossible because his aircraft were of different types, thus there was a great diversity in the speed at which each flew. Kitchener's retort was merely to reiterate: "When I come down to inspect you, you will have four machines paraded for me, to fly past in formation." Accordingly, in danger of collision and raggedly spaced, some aircraft did straggle and stagger through the air for his benefit.

The USA was regarding the war from a distance but not with total detachment. Many Americans, mostly through Canada, were making their way to Britain to join the RFC. Some who were already in Europe on the outbreak of war were enrolling in l'Aviation Militaire. Their own country, so much admired everywhere for its enterprise, wealth and vitality, had lagged far behind the warring nations. This was all the more unexpected because as early as 1907 an Aeronautical Division had been set up in the Signals Corps. At first it had balloons only, but in 1908 it received a small dirigible airship, and, in 1909, an aeroplane. It did not acquire a second aeroplane until 1911; lent by a civilian. In 1913 Glen Curtiss opened his aviation school at North Island, San Diego, and the Signals Corps Aviation School at College Park, Maryland had started with twenty-eight aeroplanes. When hostilities in Europe began, nine of the latter had been wrecked and ten of the forty pupils killed. The US Aeronautical Division owned only twenty aeroplanes. Congress authorised an establishment of 60 officers and 260 other ranks, but allocated too little money to enable a force of this size to be formed.

Of the Americans who volunteered to fight on the Allied side before their country entered the war, Raoul Gervais Victor Lufbery was the first. Life had not been easy for Lufbery. His forebears had emigrated from England to America in the eighteenth century. His father, an industrial chemist, settled in France in 1876 and married a French-woman. "Luf" was born there on 14th March 1885. Six years later his father abandoned wife and child. The boy had to start work at the age of twelve. By the time he was twenty-seven he had done casual jobs in Algeria, Tunisia, Egypt, Turkey, Rumania, Austria, Germany, Mexico, Canada and the USA. He found himself in Cairo shortly before his twenty-first birthday; and due, on his majority, to be conscripted in the French Army. He told the French Consul that he was taking his father's American citizenship, and continued his travels. He served three years in the United States Army, then went to India. There he became mechanic to the pioneer French pilot, Marc Pourpe, who started teaching him to fly.

At the outbreak of war Lufbery and Pourpe were back in France to buy a new Morane-Saulnier for their next exhibition tour. The French ignored Lufbery's claim to American nationality and his voluntary service in the American Army from 1908 to 1911, and called him up for the Foreign Legion. Pourpe had at once volunteered for the Air Service and joined Escadrille MS23. It was not long before he was able to have Lufbery transferred as his mechanic but was shot down and killed behind enemy lines on 2nd December 1914, whereupon Lufbery applied to remuster to pilot. At the turn of the year 1914 he was awaiting a training course. He began it in May and qualified in July.

The man who would emerge from this war as America's foremost fighter ace, Edward Rickenbacker, was another who knew hardship in his boyhood. He also was twelve years old when, in 1902, his father died and he had to start work as a garage mechanic. Through a correspondence course, he qualified as an engineer. By 1908 he was road testing, for the Frayer-Miller company, cars whose sales appeal lay in their high speed. He took up race driving, set a record of 138 m.p.h. at Indianapolis in a Blitzen Benz, and, by 1915, was famous and earning $40,000 a year.

Raymond Collishaw, the Canadian who, at the end of the war, would be the RAF's third highest scoring fighter pilot, was to spend most of

1915 trying to get himself taught to fly and shipped to England. Early in the year he heard that the RNAS was recruiting pilots. He resigned from the Canadian Fisheries Protection Service and joined the Royal Canadian Navy as a temporary probationary sub-lieutenant. Before being embodied, he had to obtain a pilot's certificate at his own expense. He paid $400, equivalent to £75, for a course at the Curtiss Flying Training School in Toronto, where the Chief Instructor's method of teaching – there was no intercommunication by speaking tube – was to bellow "Steer, damn you, steer," at his pupils above the roar of the engine and the howling of the wind in the struts and wires. It is not surprising that autumn gave way to winter, and weather unfit for flying, before "Collie" was able to take his test.

He was one of several volunteers who had given up their jobs and were short of money. They sent a deputation to Ottawa to ask the Defence Minister, Tom Hughes, for financial help. The advice Hughes gave the deputation leader was less than encouraging and, coming from such a quarter, astounding. "My poor boy. You and your friends have indeed been led astray and I am sorry for you. I cannot see what possible use the aeroplane is in this war. If I were the commander of a force in the field and I wished to see what the enemy was doing, I should climb a hill. If the hill was not high enough, I should climb a tree on the hill. The aeroplane is an invention of the devil and all that it has done is to draw away from us many of our best young men. My advice to you and your friends, young man, is to forget all about it and join the infantry."

The War Office and Admiralty in London rescued them from their plight and agreed to ship all accepted candidates to England, whether or not they had obtained their certificates. They were offered three choices: to continue their training at a flying school in the USA; to go home and wait until their turns came to be commissioned and sent to England; or to join a special company of the Royal Canadian Navy Volunteer Reserve that was being formed aboard HMCS *Niobe* at Halifax, Nova Scotia. Eighteen, Collishaw among them, chose the last option and he was made petty officer in charge of a batch of five. His fitness for command was already apparent. On arrival at *Niobe* he reverted to able seaman: a useful warning of the vicissitudes of life in the armed forces. They were still undergoing training in drill, signalling, armament and general Naval knowledge when 1915 ended.

To the future Marshal of the Royal Air Force and Chairman of British European Airways, Lord Douglas of Kirtleside, life had been gentler than to those American and Canadian contemporaries. In 1915 he was already embarked on a brilliant flying career.

At Oxford University, in 1913, William Sholto Douglas had joined the Officers' Training Corps as a driver in the artillery. He volunteered for service when war broke out, was commissioned in the Royal Artillery on 15th August 1914, and posted to the Royal Horse Artillery. By November he was in France. He had once seen aeroplanes at Farnborough. Now the daily sight of them overhead attracted him to the idea of flying.

Shortly before the end of 1914, it was decided that the haphazard use of other pilots and of ground crew part-time volunteers as observers was not good enough. An important part of an observer's job was artillery spotting, so transfers from the artillery were invited. Douglas immediately applied. On 26th December 1914 he was detached to No. 2 Squadron for three weeks' trial.

After the rigid discipline of the RHA, he revelled in the free and easy atmosphere of the Flying Corps. "The RFC and RAF," he wrote in old age, "offered encouragement to a certain quality of individuality. Flying attracted men whose outlook on life was tinged with an exceptional independence." That he had an abundance of the qualities which made him one of them is apparent from a report of his a couple of years later, as a pilot. The mastery of a spin had only recently been discovered. Sholto Douglas wanted to know if an *inverted* spin could also be corrected. He climbed to 5000 feet, turned his aircraft upside down and "put it into a spin to see what would happen".

For the present, however, Second Lieutenant Douglas was merely a novice observer. Lamp signalling from air to ground had been introduced, so he had to learn Morse. On his very first flight he was sent over the lines and was dismayed to find that he could not recognise any ground features at all. His pilot had to write the reconnaissance report for him. Soon he was flying almost daily and often twice a day. When he mentioned having owned a camera as a boy, he was made squadron photographer. To take photographs, he had to cut a hole in the floor of his cockpit. The cumbersome apparatus, the icy buffeting wind and his frozen hands combined to cause many spoiled plates.

Because he was hefty, he and his pilot, to save weight, did not carry rifles. Flying with Harvey-Kelly, he saw an enemy aircraft at close quarters for the first time and was able only to exchange waves with the Germans, instead of shooting them down, as an observer named Lascelles, on another squadron, had done. Henceforth Douglas always flew with a carbine.

He remained with the squadron beyond his scheduled three weeks, and, after six weeks, was told that his transfer had been approved. He could now put up the new observer's badge, for which one qualified by making a flight over the Front: a letter "O" with a single wing sprouting from it, in white silk thread. Known at once and ever after as "the flying arsehole", it remained the observers' badge until 1942, when it was replaced by an "N" for navigator, in a laurel wreath with a single wing.

He had applied for pilot training and on 26th May was sent to a small flying school at Crotoy that the RFC had opened for converting observers to pilots, on Caudrons. In his first week he made thirteen dual-control flights – six in one day – totalling four hours and twenty minutes. On 2nd June he made his first solo, of ten minutes, did another ten minutes solo, then was subjected to an hour's test during which he had to do a figure of eight and several landings on a marked spot.

Having gained his Royal Aero Club certificate in seven days, he was sent to England to complete his training to wings standard at Shoreham, on Maurice Farman Shorthorns and Longhorns, and Caudrons. The Flying Corps had quickly earned a reputation for wildness, and on one of several boisterous nights in London he and others on the course stole an ambulance and parked it outside the Piccadilly Hotel. The Provost Marshal demanded retribution, so the lieutenant colonel commanding the Training Wing had the Training Squadron moved to Gosport where he could keep an eye on them.

With a total of twenty-five hours, Douglas received his pilot's brevet. By the time he was posted to No. 8 Squadron, in France, he had increased this to thirty-two hours and fifteen minutes. Given a week's leave, he preferred to spend it at Gosport and acquire further flying time. When he reached his new squadron on 18th August he had logged forty hours and five minutes. The average pilot at that time had about twenty hours when he joined a squadron: which Douglas called "sheer murder".

James McCudden, making his way through the ranks towards the pilot's wings which it would take him some three years' service to win, was having a much tougher time. The first three months in France had offered him few comforts. In the chilly late October weather of 1914 the ground crews were still sleeping under the aircraft. He had been unable to change his underclothes for nine weeks. At the same time, he had his second bath since arriving at the Front: in rainwater collected in the folds of a canvas hangar.

Some of the Farman F20s had been fitted with Lewis guns. No. 3 Squadron had two of them. Neither proved capable of overhauling an enemy aeroplane, let alone shooting it down. Then the Squadron acquired a Bristol Scout and an SE2, on both of which rifles were mounted on either side, pointing at an angle of 45 degrees, to avoid the propeller. Neither of these machines accounted for an enemy, either.

The squadron was already, in those first few months, using a procedure that became customary in the Second World War: working from an advanced aerodrome closer to the fighting than its home base. From St Omer they flew to Gorre for each day's operations; and immediately came under heavy German artillery fire, which McCudden described as "quite uncomfortable". In addition to doing his job as a mechanic, he flew often as observer for Lieutenant Conran, armed with a rifle. On 20th November 1914 he was promoted to corporal.

Near the end of November, by when the first snow had fallen, the squadron was posted to Gonnehem, where a beet field had to be converted to an aerodrome. An Indian cavalry regiment was trying to level the ground with a roller. Everyone on the squadron who could be spared was fallen in and marched up and down, stamping in the beet and hardening the soil. Strong winds and torrential rain kept flattening the wood-framed, canvas-covered Bessoneaux hangars. The men had dug drainage ditches around these, "and every minute or so one heard a loud splash to the accompaniment of curses and oaths as some unfortunate mechanic fell into one". The rain-soaked Gnome engines could barely develop enough power to lift aircraft off the sticky ground, and soon twelve tons of cinders were delivered daily to be spread, so that aircraft could take off more easily and land without digging their wheels in and tipping onto their noses. From here, again, a forward aerodrome at Fosse was used. Because of the difficulty in taking off, the pilots used to fly there alone. McCudden wrote: "I used to enjoy these trips very much as we had a nice five-mile car ride."

In February 1915 the squadron was experimenting with night flying. Conran and Pretyman went up on the first night and landed successfully without flares. Another notable occasion was the bombing of Brussels, in which all squadrons took part. From No. 3, Birch went carrying six twenty-pounders. He returned three weeks later, having landed with engine trouble in neutral Holland, whence he escaped disguised as a ship's fireman: which makes one wonder what the distinctive garb could have been that would identify him thus. It recalls the contemporary music hall song "My old man's a fireman on the Elder and Dempster Line ... he wears a bloomin' muffler around his bloomin' throat ..." It seems rather a thin disguise.

Excitement, danger and calls for a display of heroism were not found only in the air. One incident was of a kind that became all too familiar on bomber stations a quarter of a century later. One March evening, McCudden passed Captain Cholmondeley outside the A Flight shed, where his Morane was being bombed. A few minutes later he heard two explosions and felt their shock waves. The Morane was on fire. He ran to help and found eleven dead men – including Cholmondeley, "one of the best liked officers and best pilots" – and two badly mutilated lying around it. With four more bombs likely to explode at any moment, McCudden and the other rescuers swiftly moved the victims. Major Salmond arrived, ordered everyone else away and remained there: "A splendid example of coolness that still further increased our respect for our Commanding Officer."

One May morning, Lieutenant Corbett-Wilson and Captain Woodiwiss did not return from a reconnaissance. A German dropped a note on the aerodrome to say that anti-aircraft guns had shot them down over Fournes and they had been buried there.

In July, McCudden applied for a pilots' course but was held back to continue acting as an observer and doing his mechanic's job. In compensation, he was given a test to qualify as an official trainee observer: a rôle he had been playing unofficially for several months without the reservation of "trainee". "I was given a map and told to direct my pilot, Captain Harvey-Kelly, DSO, C Flight Commander, where to go. That evening we left the ground at about six p.m. and the course I had to direct the pilot to fly over was Béthune, Lillers, Aire, Hazebrouck, Cassel, Armentières, Merville, Béthune and home. About seven-thirty p.m. we arrived back and I was very pleased and proud at having successfully accomplished the task which now qualified me to be trained as an observer." So, although he had been flying frequently over the Front for months, he was not yet entitled to wear the badge that Sholto Douglas had been awarded in a far shorter time and after one flight over the trenches.

In November Major Ludlow-Hewitt took over the squadron, which another officer destined for high rank, Lieutenant Portal, had recently joined as an observer.*

In the intense cold of winter McCudden suffered a frostbitten face, so took to wearing one of the protective masks that were issued. Another time he had a severe headache all day from flying at 11,000 feet, the highest he had yet been.

He ended 1915 as a sergeant, still waiting for his pilot's course.

Among the French, one of the most colourful, eccentric and reckless characters in the history of aviation was becoming conspicuous in a manner that contrasted with the conventional and conscientious McCudden's, and with other, more serious, Frenchmen such as Garros and Guynemer, as dramatically as a peacock with starlings.

Charles Nungesser was twenty-two when he enlisted in the hussars

* Both were future Commanders-in-Chief of Bomber Command and retired as Air Chief Marshal Sir Edgar Ludlow-Hewitt and Marshal of the Royal Air Force Lord Portal, respectively.

in August 1914. He had crammed more variety, enterprise and risk into his years than orthodox men experience in thrice as many.

Like Sholto Douglas, he was of middle height, burly and good-looking. Both were good athletes, Douglas a rugger player, Nungesser a horseman, boxer and swimming champion. He raced cars and motorcycles: speed and breakneck hazards were irresistible to him. He, in turn, was notoriously as attractive to women. Learning to fly, he went solo immediately. That was not unusual in those days. Louis Strange, describing how he was taught at Hendon – by a Frenchman, Louis Noel – said: "We knew that the machines into which they put us would fly, and we had expert instructors who could *tell* us how to fly them. All we had to do was to obey the instructor's instructions and fly."

Strange went on to recount how Monsieur Noel despatched his pupils on their maiden attempts: "I have told you how to fly. You have understood, yes? Very well, I give you the last chance to say 'no'. Very well, you can fly, do you hear? I, Louis Noel, say you can fly. I speak no more. I go to the bar. If you commit suicide, that is bad; but if you *almost* do that, it will be *much, much worse* for you."

Soon after Nungesser had survived a similar robust empirical indoctrination, his family suffered a financial misfortune. He went to Brazil to find a rich uncle who had departed from France without leaving an address. Uncle rediscovered, the dashing nephew stayed a while to try planting coffee before returning home to join a compatriot in organising flying meetings.

Although he went into the cavalry, l'Aviation Militaire was in his mind. On 20th August 1914 he was at the Front. On 3rd September he performed a feat of arms that was hardly less astonishing than Horatius's holding the bridge. He, a mere trooper, a cavalry lieutenant and two infantry privates were cut off in an enemy advance. The officer, wounded, hid. The other three lay in a ditch, and, when a powerful Mors car appeared at speed, Nungesser stopped it by closing a level-crossing gate; then, from ambush, he and his two companions shot all four occupants, and Nungesser drove the Mors furiously back to his regimental Headquarters. This episode is usually described in a way which implies that Nungesser disposed of the four Germans single-handed. It is common sense that this would have been impossible without hand grenades or a machinegun, either of which would have wrecked the car. The attack, anyway, needs no exaggeration. It was an act of outstanding bravery as it stands and Nungesser displayed initiative and leadership as well as courage. He was allowed to keep the bullet-holed car, promoted to corporal (two months before McCudden) and decorated with the

Médaille Militaire. He became known as "l'Hussard de Mors", but with characteristic mordant humour rechristened himself "l'Hussard de Mort".

In January 1915 he was appointed a driver at 33 Corps HQ. What aberrant altruism prompted any high-ranking officer to commit himself to the mercies of so reckless a chauffeur is hard to fathom. He now transferred to the Air Service and qualified as a military pilot on 2nd March. On 8th April he was posted to Escadrille V106 and lost no time in distinguishing himself on bombing raids. On the 22nd April he received his first citation for his part in operations on the 15th, 16th and 17th, and was promoted to sergeant: which, in the French cavalry, bears the resounding title of "Maréchal des Logis".

His mechanic, Pochon, was also his friend and used to drive him back to camp while he slept on the rear seat, after a night's roistering in town. By 15th May, Nungesser had flown fifty-three day and night operations and was made a warrant officer. On 31st July he shot down his first enemy aircraft: one of five, which he hit in its carburettor and forced to land.

In November he was posted to a fighter escadrille, N65, with which he stayed until the end of the war. He celebrated his arrival by weaving among the chimneypots and steeples of Nancy, looping over the main square and flying along the main street at thirty feet. Undeterred by his CO's reproof, it was not long before, on 26th, the forcefulness of his personality prevailed on some of his comrades to accompany him to an enemy airfield where, while they kept guard, he gave a spectacular performance of aerobatics and strafing: a "beat-up". For this he spent eight days in open arrest. Two days later, while still under arrest and ostensibly on an air test, he again shot down an enemy aeroplane.

At the end of 1915 he was still awaiting a commission.

Jean Marie Dominique Navarre was as ebullient a pilot as Nungesser and of no less worth to their country. Although his score of enemy aircraft destroyed was much the smaller, it was he whom the archives credit with having been the great innovator. "At Verdun," the records say, "Navarre innovated combat between aircraft, and methodical attack. He demonstrated how the little monoplane, obedient to one sole will, could become a dangerous weapon for its adversary, in the hands of a pilot who was skilful, experienced and brave to excess."

The Battle of Verdun still lay a year and a half in the future when Navarre joined l'Aviation Militaire on war's outbreak five days after his nineteenth birthday. He had been learning to fly at Crotoy, the same school where Sholto Douglas would soon be converting from observer to pilot, but interrupted his training to go to the colours. He obtained his military pilot's brevet the following month. The records say of him: "It is true that, like Garros, Navarre possessed a bird's sense of flight." He was posted, a second lieutenant, to bombers, Escadrille MF8, where "He revealed extraordinary coolness." He was posted to N12, to fly fighters, a year later. The end of 1915 found him with two victories to his credit, patiently and methodically learning his new craft.

On the other side of the lines, the aeroplane pilots and observers of the German Military Aviation Service were diligently filling the gaps in their professional knowledge that were a result of their country's obsession with airships.

Among the names on which fortune has conferred fame in air fighting, Max Immelmann's was one of the first. His brother describes him as a studious boy who was calm and thoughtful, affable and modest, self-assured and self-reliant; and comes dangerously near to making him sound a prig. He was certainly ascetic. As a cadet he was chaffed by his comrades for his dislike of meat and abstention from alcohol.

In 1905, aged fifteen, he entered the Cadet School at Dresden, where he found it difficult to adapt to the discipline and stiff etiquette. On 4th April 1911, the youngest ensign in the German Army, he joined the Second Railway Regiment. This hardly sounds like a lively environment, and his resignation the following year, to study at Dresden's Technical High School, causes no surprise.

On 10th August 1914 he saw a notice asking for volunteers for the Aviation Corps, and applied. Two days later he was mobilised and it was not until 12th November that he was posted to the Aviation Replacements Section at Adlershof, from where pilots and observers were sent on courses at various aircraft factories. He learned to fly at LVG, where he was also taught about engines, aeroplane construction and meteorology and how to use a compass. On his eighth day he began flying with several ascents at heights of fifty to eighty metres, and landings, under an instructor. After fifty-four flights he went solo on 31st January 1915. On 9th February he passed the pilots' test, which demanded five figures of eight and a landing after each at a place marked by a red flag. Then came the preliminary test for a "war pilot": 20 smooth landings and two flights of 30 minutes at 500 metres. After that he had to pass the actual "war pilot's" test of one hour at 2000 metres and a glide down from 800 metres. Next, on 11th February, came the "field pilot's" test. On this, he had to climb to 2600 metres, which took him 65 minutes, then descend to 2400 metres for 20 minutes' straight and level flying, followed by another short drop to 2200, from where he glided down to land in three minutes. His instructor was very pleased.

He was naïf, it seems. He wrote to tell his mother that he was very popular with his superiors, his fellow pupils and his subordinates. Was he clairvoyant, that he could divine what underlings thought of him? They would hardly have fawned on him with expressions of praise.

On 12th February he returned to the Aviation Replacements Section for further instruction. By the end of the month he had made forty-five flights, but none across country, and was assessed the best pilot on the course. Two pilots were wanted for the Front and Immelmann was told he could be one of them if he demonstrated a landing from 800 metres and another landing on rough ground. Attempting the latter, he hit a manure heap and turned over, was unhurt but failed to qualify for the front line. Finally he had to make fifteen landings outside the aerodrome and do three cross-country flights.

On 12th April he joined No. 10 Field Section for artillery co-operation. His aeroplane was prepared for operations by the fitting of metal sheets under the fuel tank and seats, a bomb rack and map board.

There were ten officers on the section, five pilots and five observers. "The gentlemen are all very nice," he wrote to his mother, "but that is a matter of course with airmen."

He spent only two weeks there before being transferred to Flying Section 62, which was being formed under the command of Hauptmann Kastner to fly LVGs. Three of the six pilots were lance corporals but all six observers were officers; four of them ex-cavalrymen. One of the pilots was Oswald Boelcke, with whom Immelmann formed a close friendship. They had in common a serious disposition, the constant search for innovations that would make them and their Abteilung more efficient, and reserved, highly self-disciplined natures. Boelcke, a devout Catholic and in every sense a good man, evolved into the first great fighter leader and remains one of the most admired figures in the development of air fighting.

In May 1915, Flying Section 62 was among those which were re-equipped with the new Fokker E1, a single-seater monoplane that revolutionised the situation in the air over the Western Front by devastating the French and British air forces. At the year's end, Immelmann had shot down five aircraft and Boelcke three: with many more in prospect for them both.

More typical of the average pilot, who did not attain fame but shared all the dangers of those who did, was Hauptmann A. D. Haupt-Heydemarck. He at once gives the impression of having been an easy-going, good-natured man, a pleasant companion. Initially, he was not fired with any particular zeal for flying; he went in for it because the chance came his way and he took it more out of curiosity and boredom than with dedication. Patience he certainly had: his flying apprenticeship began in the summer of 1912 and, after a two-year interim, was resumed in 1915. He was an infantry lieutenant when his regimental Adjutant announced that volunteers were needed for training as pilots and observers. Every officer stepped forward, but the Army was in no hurry and they were left waiting months for a summons to flying school.

One late autumn day an offer came from the Chemnitz Flying Club for two of them, as a consolation, to have a preliminary taste of aviation, on a balloon flight. Lots were drawn. "In compensation for my recent bad luck in love," Haupt-Heydemarck writes, "I was lucky in gambling and won." The ascent was set for All Saints' Day, with Professor Beuermann, of the Chemnitz Technical Institute, as pilot; but when the two passengers reported to him, he had to disappoint them. "Unfortunately we can't take off today, gentlemen: we have an eighty kilometres an hour wind."

"So we'll fly all the faster," suggested Haupt-Heydemarck.

"And the landing?" the Professor said.

"I kept pressing him," and eventually, from a sheltered place, up they went. The lift-off was pleasant. The wind was strong. "I felt myself the Lord of the Air and was content to have had my way. Landing? Oh, Beuermann was an experienced pilot and would bring us down safely!" The strong wind bore them eastwards. "After two hours of hedonistic enjoyment to the full, we met snow clouds." The Professor released some gas and down they went until the ground was in sight again. "The view was indeed not gratifying: we were being wafted towards a chain of three long lakes.

"'We must land at once,' Beuermann said, 'or we'll have a horrifying end in this bitter cold.'" He does not sound the optimistic sort who would be one's first choice as companion on a risky venture, but H-H seems not to have been much perturbed. Most people would probably have had reservations in the first place about a pilot who allowed his passenger to talk him into going up against his better judgment of weather conditions; but this passenger, as will be seen, was preternaturally cool in temperament.

They descended rapidly. "From a height of 1000 metres, the ground had seemed to slide past beneath us slowly, but now that we were so low over them, the trees were rushing by. The storm had not abated. At a speed of 80 k.p.h., ought Beuermann to put the basket down on the frozen ground? That could result in matchwood and splintered bones. Last instruction: 'On landing, bend your knees deeply!'

"What came now happened so quickly that it was almost too swift to comprehend. One more piece of luck, that the ground was fairly level.

The first lake was looming critically closer. Beuermann energetically pulled the ripcord. It tore a great panel out of the balloon, so that the gas could escape fast. With no buoyancy, the balloon should touch down in a moment. Unfortunately, things did not work out according to plan. The high wind caught the empty envelope and drove it upwards house-high. Then we were dragged down in its damp folds.

"Alarmingly, we rushed earthwards – now the aforementioned knees bend – 'Rumps!' cracked the basket. Its broken pieces were scattered over the frozen soil. It turned over and by good fortune stopped.

"We were catapulted out and lay bruised and battered. Beuermann had hit his head on the basket's hard rim and was wiping blood from his ankle. I tried painfully to stand up but my left leg hung limp and I couldn't. An unpleasant certainty grew on me: a broken thigh! With a resigned laugh I lay down again. Swinish luck!

"My comrades put a splint on my leg and bedded me down on a straw-packed farm cart. In a mild snowfall the wagonload jolted off on its unpleasant way to Gitschin. There I was given medical attention and on the next morning went back to Chemnitz by train. At the railway station I was treated with curiosity and respect; rumours buzzed: 'Officer wounded in a duel!'

"In the garrison hospital a medical officer put on a grave face: my right leg had shrunk a full ten centimetres! Uniform, farewell!

"But his skill stretched the broken bones apart so that finally a shortening of only three centimetres remained. So, by slouching a bit on the other hip, I was able to remain a soldier.

"Three months later I made my second balloon flight: this time with a smooth landing."

In August 1914 he was sent as Adjutant to a brigade at the Front. He says that his prospects of flying hung by a thread, but he did not give up hope. In the summer of 1915 he was called to a short course at the Aviation Replacements Section. There were several fatal accidents and he would be happy to return to the Front. When 1915 came to an end he was still wondering to what type of Flying Section he would be sent: artillery co-operation; a Battle Section that dropped bombs; a Corps Section; or one that did long-range reconnaissance?

When making a roll call of great airmen, the names Heinrich (Heini) Gontermann, Ritter Fritz von Röth and Leo Leonhardn do not spring at once to mind. Yet all were eminent enough to win their country's highest decoration, the "Blue Max", the Pour le Mérite. Why a designation in the language of Germany's most hated enemy? Because Frederick the Great, King of Prussia in 1740–86, who instituted the decoration, could speak only French.

Von Röth and Gontermann were famous for destroying balloons: balloon-busters, as the RFC called the specialists in this dangerous expertise. Von Röth was known as "die Fesselballon-Kanone", "the Captive Balloon Cannon", and "Cannon" was the name the Germans gave to their "aces". Gontermann was dubbed "der Ballonkiller". But at the beginning of 1915 observation balloons were still scarce and the British had not begun to use them at all. Gontermann began the war as an eighteen-year-old officer of lancers and went to the Front on 13th September. The year 1915 saw two changes of direction in his military career. In June he was sent to a machinegun school and thence to the machinegun company of the 80th Fusilier Regiment. In November, he began the pilot's course for which he had applied. Von Röth was an artillery officer, twenty-one when war was declared. His regiment went to the Front immediately and on 10th September he was gravely wounded in the head and a lung. During his long spell in hospital he heard fhat he would be released from the Service. The anxiety that his part in the war was over haunted him, and as soon as he was discharged from hospital he volunteered for flying duties. Towards the end of 1915 he began his pilot training.

Leonhardn is particularly interesting because of his age. Born on 13th November 1880 in East Prussia, he was on the threshold of his middle years when the war began. Photographed four years later, he looks the Briton's mental image of a typical Prussian: square-skulled, balding, close-cropped, barrel-chested. But there is no cruelty about the set of his lips; his expression is quite cordial: perhaps from pride at the newly awarded Blue Max at his throat and the medals on his chest. He was a hard man, known as "der eisener Kommandeur", "the Iron Commanding Officer".

At nineteen he was an ensign in the Pomeranian Fusiliers. From 1908 to 1912 he was Adjutant of the Queen Victoria of Sweden Regiment, then spent a year with the 138th Infantry before going on a pilot's course on 1st February 1914. He must have received a rude shock nine days later, for it is on his records that on 10th February he executed an involuntary loop and "his year-long desire to join the Air Service almost came to a premature end".

The record goes on: "A collision with an aeroplane from another flying school nearly caused Leonhardn to quit this life. Coming in to land from the opposite direction, it caught the propeller of his Taube, which somersaulted several times and hurled him to the ground."

Leonhardn himself wrote: "With this I established an unusual world record: I broke my spine in two places, breastbone, nose, the base of my skull again, suffered concussion, lung and liver ruptures, and crushed my left knee so that the bones splintered badly. Generally, nobody survived such injuries!" No wonder he wears a hint of a smirk in his photograph.

The will to live and fly soon put him back on his feet. He was in hospital in Berlin until 7th June and then in Wiesbaden, where his promotion to Hauptmann came through. On 5th December he left his wheelchair in Wiesbaden to become Adjutant at the Flying Units Inspectorate in Berlin.

"By careful avoidance of doctors," he says, with a dry humour that prompts one's liking as well as admiration for his fortitude, "I contrived to get away from there and back into the war." He was posted as Adjutant to the Southern Army's Aircraft Park. Thence, on 1st May 1915, he went to Flying Section 59 as an observer. On 13th August he took command of Section 25.

The Italian nation knew from the moment war broke out on two Fronts that it would not for long remain neutral. On the Western Front, Germany was soon in stalemate with France and Britain. On the Eastern Front, where Germany and Austria–Hungary faced Russia, and distances were far greater than on the west, a conventional war of mobility and fluid lines was being fought. Where opposing forces did establish trench lines, these were temporary.

Austria was Italy's traditional enemy. The city of Trieste and the district of Trentino had once been Italian but had become Austrian centuries earlier. This cut off nearly half a million Italians from their homeland and was a smouldering grievance.

The right man to stoke the embers and stir them into flame was ready and waiting to do just that. Gabriele d'Annunzio, born in Pescara in 1863, was by no means too old at fifty-one, when Germany marched against the Allies of the Entente Cordiale, to be actively belligerent. A fiery little man with a scrawny physique and plain features, he wore a moustache whose upward-curving ends gave him a comical, even clownish, rather than aggressive, look. His appearance did not suggest an esteemed writer and poet, a duellist or notorious lover: all of which he was.

A fervent – and verbose – patriot, at the age of thirteen he had written to one of his school masters: "My first mission on this earth is to teach the people to love their country and to be honest citizens. The second is to hate to death the enemies of Italy and to fight them always."

Due for military service in 1883 when at university, he obtained a six-year postponement; then, in 1889–90, he served as a Reserve officer in the 14th Cavalry Regiment (the Novara Lancers), stationed in Rome.

Flying interested him from its inception and in 1909 he attended Italy's first important international air display and competition, at Brescia. Many of the great pioneers of aviation took part: Wilbur Wright, Glenn Curtiss, Louis Blériot, Henri Rougier, Alfred Leblanc; and, among the Italians, Mario Calderara and Umberto Cagni. The "Gran Premio Città di Brescia", for five circuits of the ten-kilometre course, was won by Curtiss in 51 minutes and 52 seconds; with Calderara second and Rougier third.

D'Annunzio asked Curtiss to take him up. The American did not usually grant this request, because of the risk involved; and at Brescia there had been some accidents, one fatal. But, for d'Annunzio, he consented and gave him an eight-minute flight. The next day, d'Annunzio flew with Calderara.

The official account of his military flying career in the archives of the Italian General Staff says: "The great poets, if they are real genii – and d'Annunzio was – have also the faculty of prophecy. The fact is that the Pescaran poet recognised the inevitable revolutionary importance that the conquest of the air would have on the destiny of the world, in peace and war. And at once propagated his convictions." On 21st February 1910 he held his first conference on aviation, at the Lirico Theatre in Milan, at which he "recounted the short but extraordinary story of the conquest of the air, starting with Wright's first flight on 17th December 1903".

In his discourse, d'Annunzio urged the development of flying and recommended that prizes should be offered for flights from Milan as far as Genoa, Turin, the peaks of Generoso and Mottarone.

"We are celebrating today," he added, "a game of audacity; we are on the eve of a great change in social life. The code of the air is being established. The frontier invades the clouds."

He was among the first and the few who saw the flying machine as a new arm that would influence warfare with increasing decisiveness until it became the determining factor. He would, in the future, express this intuition of his by originating for the bomber squadrons the proud motto *"Suis Viribus Pollens"*, *"possente di sua forza sola"* (Powerful in its unique might).

"This," the archives say, "because he understood that the air weapon lent itself more than any other arm to multiplying and extending offensive power immeasurably, more than guns, more than ships; so that, when the authorities of the great nations were determined on the supreme importance of their navies, he dared to say [marrying Greek to Latin]: 'Uranocrazia is about to replace thalassocrazia: that is, naval supremacy is about to give way to aerial supremacy.'"

[63]

In March 1912 the National Air League was founded in Italy, with the purpose of diffusing a knowledge of flying throughout the country. D'Annunzio, in voluntary exile in France, hastened to give his support, extolling the initiative with a telegram of good wishes: "May the Naval League and the Air League be the two indefatigable arms of the new power."

On the outbreak of war in 1914 d'Annunzio returned from France. In Genoa on 5th March 1915 he gave an inflammatory interventionist speech. Italy declared war on Austria on 24th May. Fifty-two years old, he voluntarily rejoined the Army.

General Luigi Cadorna, Chief of the General Staff of the Italian Army, wrote to d'Annunzio on 25th May: "Illustrious Sir, I have received with great pleasure the expression of your wish to participate directly in the enterprise to whose preparation you have brought a contribution of high ideals. I have therefore interested His Excellency the Minister of War, so that he may take steps to recall you for service as an officer in the Novara Lancers."

The general also arranged for d'Annunzio to be posted to the Army Headquarters commanded by His Royal Highness the Duke of Aosta. There, he would be able "to render a valuable contribution both practical and advisory". He would also be authorised to visit all the other Army Headquarters "to witness the events which will occur on the whole Front".

This effusive and flattering reception of d'Annunzio's return to uniform did not offer what he sought. "I did not expect propaganda work of a journalistic nature," he said. He wanted to fight. He accepted the appointment, but, on 30th July, wrote to the Head of the Government, Antonio Salandra:

"You know with what impatience I have requested the honour of serving my country in other fields. And for your solicitous kindness in seconding my desire, I have not yet ceased to give thanks.

"You know that I have awaited this hour all my life. Having lived with sadness and wrath among a people careless of glory, at last I witnessed a miracle, which responded to my implacable expectations. Glory has become the very sky of Italy. The hour of great deeds has sounded for this nation, the hour of my blood has come for me.

"I arranged, with the most valorous Lieutenant Giuseppe Miraglia, an enterprise to Trieste. Experienced in aviation, having already flown many times and at high altitude, being gifted with a certain ability for observation, and knowing the topography of Istria, especially the layout of Trieste from my numerous visits, I was thinking of the usefulness and the beauty of a flight which would carry a message to the tortured city and, possibly, some damage to military installations adjacent to Santa Teresa dock.

"Everything was prepared with the severest discipline. The probability of success was great. The aeroplane could climb higher than three thousand metres ..." But a newspaper heard of the enterprise and not only compromised it but also "provoked various remonstrances from the Ministry". Whereupon d'Annunzio was forbidden to take part in "so dangerous a venture. I cannot tell you how saddened, stupefied and offended I was."

Although he romanticised warfare and his own part in it, in the manner of one who has never been in action, so does not know the reality of war and its horrors, and longed to be acclaimed a hero, he was sincere in his wish to fight. "How is it possible in my regard to speak of 'a precious life' and 'a duty not to expose myself'? I am not a literary man of the old sort in a skullcap and slippers. It is perhaps easier to restrain the wind than to restrain me. I am a soldier, have wished to be a soldier, not to spend my time in cafés and the mess, but to do what soldiers do."

On 7th August 1915 d'Annunzio was authorised to fly on operations.

CHAPTER 6

1915. In Action

As the pilots of both sides shared many attitudes and felt a mutual understanding while they set about doing their utmost to kill each other, so the most senior officers who commanded or otherwise influenced the air forces had, among their many disparities of character, one quality in common: an early appreciation of the importance of air power.

It was because of the intimacy created by the RFC's small size that Henderson, as its Commander-in-Chief, and Trenchard, commanding a wing, had an advantage over those who held corresponding offices in the French and German air forces. Among the Germans and the French there were colonels and generals whose existence the flying men registered but did not entirely notice. With the British, it was different. Henderson's and Trenchard's frequent visits to the squadrons made everyone aware of who led them at the summit. They well knew Henderson's gentleness and compassion, his total imperturbability and subtle penetrating intellect. They knew equally well Trenchard's energy, the stentorian voice which had given rise to his nickname, "Boom", and the visceral trembling his presence or his orders occasioned them: especially when he succeeded Henderson. Then, like Haig, whom he so much admired, Trenchard did not let loss of life deter him. He kept them flying in machines inferior to, and in smaller numbers than, the enemy, despite a casualty rate that Henderson might not have countenanced. But, at the beginning of 1915, that black period was still to befall the squadrons.

However uninterested in aeroplanes the War Office and the highest ranking British Army officers had been three or four years earlier, many, as has been seen, were succumbing to the Flying Corps' demonstration of its unique usefulness: from Grierson at the 1912 manoeuvres to Smith-Dorrien, Haig and Sir John French at the Western Front and Lord Kitchener in London.

L'Aviation Militaire was ultimately responsible to the Minister for War, who, at the outbreak, had been Messimy; replaced on 26th August 1914 by Millerand. Below him, Commander-in-Chief, was Joffre. To draw up the plans for its use, to define its needs, to decide the disposition of the fighting units – in short, to guide the war in the air – General Joffre and his successors had a specialist at their disposal in General Headquarters. Initially this was Lieutenant Colonel Voyer, who was succeeded on 26th September 1914 by Lieutenant Colonel Barès. To enable GHQ to carry out its operational task, its back-up organisation had to provide the necessary weapons and manpower. In 1914, foreseeing a short war, flying schools had been closed and the rate of aircraft production slowed. It was very quickly seen that training must be resumed and aircraft production organised.

From the archives: "The task was all the more difficult because the domain to be explored was immense and new, and the problems posed to those responsible for resolving them were complicated by frequent disagreement between the civil power and GHQ. All this explains the successive modifications in the organisation of the directorate and the changes at its head throughout the war.

"At the outbreak, General Bernard presided over the destiny of the Air Force. The mistakes made, which, however, were not all attributable to him, led to his rapid replacement as Director, on 25th October 1914, by General Hirschauer, former Inspector of Aviation, who, as soon as he took charge, restored production, reopened the old flying schools and created new ones."

Joffre, not the most imaginative of generals, had, surprisingly, shown an early awareness of the possibilities the new Arm offered. In November 1914 he wrote: "Aviation is not only, as was once supposed, an instrument of reconnaissance. It has proved, if not indispensable, at least extremely useful in the control of artillery fire. It has shown, further, by the dropping of high explosive projectiles, that it is capable of being used as an offensive weapon, either for long-range missions or in co-operation with other forces. Finally, it has, moreover, the ability to pursue and destroy enemy aeroplanes."

General Ludendorff had made his first flight four years ago and at once perceived the value of an air force. He distinguished himself on the Western Front in the first week of war. Transferred to the Eastern Front as von Hindenburg's Chief of Staff, he was the architect there of the great German victory at Tannenberg at the end of August 1914. After the battle, he declared that "without airmen, there would have been no Tannenberg". It was aerial reconnaissance, carried out by aeroplanes operating from improvised aerodromes and landing on any flat ground near the Commanding General's Headquarters, which had enabled Germany to triumph.

The organisation of its air force with which Germany entered the war had a wider span than either Britain's or France's, but was used clumsily: whereas aeroplanes are essentially instruments of swift, whiplash effect, flexible and elusive. Under the Headquarters, two separate air forces were incorporated, the Prussian and the Bavarian. The operational units consisted not only of aeroplane Abteilungen but also those that flew airships, the balloon units and the anti-aircraft batteries. Control over Flugabwehrkanonen (already known as Flak, a term not adopted by the British until the next war) was given to the air force, on the principle that anything to do with the air must be within its province; whereas in other countries the connection of anti-aircraft guns is seen as being with the artillery. Under the German system the Air Service was thus burdened with another training, as well as supply and maintenance, commitment.

The airship units were still an important part of the Service. On mobilisation, twelve dirigibles were operational. Unfortunately the great hopes that the general public had placed in airships were not fulfilled. Certainly, little by little, fifty would be in service. But airships were easy targets for anti-aircraft batteries, because their heavy bombload prevented them from flying high enough to be out of range. It was just as well that it did: the Zeppelins' main targets were in Britain, foremost among them, London.

In September 1914 a little-known, and today scarcely remembered, specialist bomber force had been formed. Consisting of two wings, each of three six-aircraft Abteilungen, it was named the General Headquarters Flying Corps, commanded by Major Wilhelm Siegert, and posted to an aerodrome near Bruges.

On the night of 28/29 January, it made the first mass air raid in history by dropping 123 bombs on Dunkirk from a height of 1000 metres. Besides being an attempt at widespread destruction of the French port, this was practice for the raids this force already planned to perpetrate on the cities of Great Britain and their civilian populations. A stroke of fortune thwarted this intention: Major Siegert's thirty-six bombers were transferred to the Eastern Front, where they did heavy damage to the Russians.

With such an unwieldy and impersonal structure and rigorous, automated Teutonic discipline, no numinous figure emerged comparable with cool cerebral Henderson, choleric Trenchard with his bravura gestures and paradoxical sensibility, or Barès, gifted with an instinctive professional appreciation of every problem and a clinical solution to it.

On 11th March 1915 the office of Chef des Feldflugwesens, Head of Field (meaning operational at the Front) Aviation was established and his title, in accordance with German military practice, abbreviated to FeldFlugchef. The man chosen is usually referred to as Major Thomsen. He was, in fact, Hermann von der Lieth-Thomsen, who at forty-eight was a most experienced pilot and as practised in wartime flying as anyone could be after so few months at war. His deputy was Major Siegert, mentioned above, who in peacetime had flown his own aeroplanes and belonged to the suspectedly barmy fraternity, of "balloonatics". "Therewith," the records claim, "at the summit of Army aviation were two men who at the right time exerted a considerable authoritative influence over the development of military flying."

From time to time, an observer armed with a rifle would bring down an enemy aircraft. On 22nd January, twelve Albatros bombers raiding Dunkirk were intercepted by a mixed handful from British, French and Belgian squadrons. One Albatros forced-landed with a bullet in the engine, fired from a Four Squadron BE2. On 5th February, a No. 3 Squadron Morane piloted by Second Lieutenant V. H. N. Wadham, with Lieutenant A. E. Borton as observer, fought for several minutes with an Aviatik, whose pilot and observer were shooting at them with pistol and rifle. After exchanges of fire at ranges from 100 yards down to 50 feet, the Aviatik was forced down.

Also on 5th February, the first Vickers Fighting Biplane, the FB5, arrived on Five Squadron and was followed by a few more. This placed the observer in the front cockpit and provided him with a Lewis gun. Its French-manufactured Gnome engine gave so much trouble that one pilot had to make twenty-two forced landings out of thirty flights, and combats were few. Machinegun stoppages were frequent, mainly caused by the intense cold at high altitude, so observers still carried rifles as well.

Anyone familiar with the history of air combat in the two World Wars might reasonably suppose that if he were to close his eyes and stick a pin into a list of pilots who have extricated themselves, at the last split second, from dire situations in which death seemed immensely more probable than survival, Louis Strange's name would exert an irresistible magnetism.

The squadrons were still receiving heterogeneous types of aeroplane. No. 6, on which Strange was now serving, was issued with, among others, a single-seater Martinsyde Scout with a Lewis gun mounted on the upper wing, so that it fired outside the propeller disc, that is, outside the radius of the blades. This meant that, to change the ammunition drum, the pilot had to loosen his safety belt and stand up. Moreover, changing the 47-round drum on the Lewis needed the exertion of more strength than might be supposed. It had to be done balancing precariously in a slipstream of 60 to 80 m.p.h.; when the 20-odd-pound weight of a full one became more than a trifle wayward to handle. Later, a 97-round one weighing 30 pounds came into use and injuries to pilots became common. Major Gordon Shepherd, the Commanding Officer, gave Strange permission to make this machine his personal mount: and the term is not as fanciful as it might seem today, because at that time pilots were said to fly "on", not "in" an aeroplane, a derivation from the horsemanship that was alleged to confer the touch essential for both.

The Martinsyde was a difficult aeroplane to fly: slow, sluggish on the controls and inherently unstable. But Strange had demonstrated above-average skill and acquired wide experience. Keen, fizzing with energy, he crossed the enemy lines to hunt for a victim and spotted an Aviatik two-seater a couple of furlongs away and a few hundred feet above, whose crew soon saw him in pursuit. Climbing, encumbered with its machinegun's weight and the resistance it offered to the air, the Martinsyde barely attained a mile a minute. It could not quite overhaul the Aviatik. Presently Strange realised that his machine would go no higher. He was at extreme range and it was now or never. Still scrabbling for a few more feet of height, he opened fire, emptied his drum, and missed.

He was twenty miles deep in Hunland, as the RFC called enemy-held territory, and limited endurance forced him to turn for home. But he might run into other enemy aircraft. He must change the ammunition drum. At 10,000 feet it was piercingly cold and lack of oxygen made exertion painfully laborious. He loosened his belt and stood. The machine, in a shallow dive, was doing about 75 m.p.h. The empty drum would not come free, however hard Strange twisted it. He wrenched. The Martinsyde dropped a wing. He lost his balance and lurched against the joystick. The machine flicked over onto its back.

Strange was hanging by his gloveless, freezing hands. One hand clung to the small round metal drum, which had stuck only because the threads were crossed. With his other hand he had grabbed the centre rear strut between the mainplanes. Beneath him was 9000 feet of empty space and the aircraft was descending. If he did not fall to his death he would be crushed to a pulp when it hit the ground.

In this position, he was doomed. He began to swing from side to side until he hooked a foot under the cockpit rim. But his pendulum motion induced a spin. No drill for recovery from a spin had yet been established. He had to find one before he lost his last 1500 feet of height. A sideways kick at the stick, a rapid slide into his seat, hands and feet on stick and rudder bar, and he corrected the aeroplane's attitude with the hedges and trees only a few feet beneath.

During the first three months of 1915 the winter weather often kept squadrons grounded. What was worse, their frail aircraft and the light, portable timber-framed canvas hangars were vulnerable in high winds. The old year had gone out with a violent storm that wrecked thirty RFC machines, sixteen of them irreparably. Reconnaissance, artillery co-operation and photography continued, but, with operations curtailed by low cloud, storms, lack of serviceable aeroplanes, there was time to spare for improving methods of communication and reorganising photographic reconnaissance.

The British, the French and the Germans were all using these weeks of patchy weather to prepare for the next head-on collision between the two sides. Spring was the time for big pushes. Land battles on a huge scale were a certainty. As time wore on, greater numbers of aeroplanes would inevitably be involved. So much was evident. The imponderables were: Who would strike first? Where? And whose aircraft would have the best performance and armament, in both of which everyone was striving for improvement?

The average flight at such a period, whether on visual or photographic reconnaissance or artillery co-operation, seldom brought contact with an enemy aircraft. If it did, the separating distance was usually too great for any attempt at aggression. The airmen might ignore each other, or acknowledge one another with a salute or a little mild showing off. If shots were fired, the great majority went wide of the target: better results could not be expected when shooting from an unsteady platform that was travelling fast, at a target that was also fast-moving, manoeuvrable and subject to sudden involuntary shifts up, down and sideways caused by air currents. The greatest danger in flying still came from accidents: on taking off or landing; in flight, through mishandling the throttle or controls or a worsening of the weather. More pilots were being killed while training than at the Front.

Of the three methods of signalling between air and ground, by Vérey light, Aldis lamps flashing the Morse code, and wireless telegraphy, the third one had the longest range and was the most efficient. The equipment, however, was so heavy that no observer could be carried; and so bulky that some of it had to be fitted in the pilot's cockpit as well as the observer's.

The accuracy with which the fall of shells was reported to the artillery batteries needed improvement. A map divided into squares identified by numbers and letters had been introduced soon after the RFC arrived in France, and where a shell landed in relation to the target was indicated by the square and "left", "right", "short" or "over". Now Captain Lewis, who had devised the square system, offered a better one. This used the clock code, with 12-o'clock dead ahead in the direction in which the guns were shooting. The observer centred on the target a celluloid disc with concentric rings marked on it at scaled distances of 10, 25, 50, 100, 200, 300 and 400 yards radius, and lettered from A to H. As each shell fell, he signalled to the gun battery the ring within which it had hit and the "time" on the clock face. This became standard.

To make the results of photography more easily available to all who needed it, Henderson sent Major Salmond to learn how the French organised this. As a result, he set up a photographic section at each Wing HQ, consisting of a pilot, an observer, and two other ranks. Lieutenant Moore-Brabazon commanded the first one, which was attached to First Wing.

No. 1 Squadron arrived at the Front on 7th March, under the command of Major W. G. H. Salmond. Equipped with eight Avro 504s and four BE8s, it joined Third Wing in time for the coming battles.

What Tennyson called "the ringing grooves of change" were about to make themselves felt. Nos 2 and 3 Squadrons' photography had given Haig clear, detailed information on the German defences opposite his First Army's front. The village of Neuve Chapelle obtruded into the British lines, was assailable on two flanks, and was made the first objective. The Battle of Neuve Chapelle began at 7.30 a.m. on 10th March with a bombardment in deteriorating weather.

For the RFC this had the special importance of being the first occasion on which planned tactical bombing was attempted as an integral part of the whole operation. Hitherto bombs had been dropped sporadically as a secondary feature of a sortie whose primary purpose was reconnaissance or artillery spotting. This time, No. 3 Squadron was ordered to bomb some buildings in Fournes reported to be a Divisional HQ. An early instance occurred here of a practice that continued in the RAF into the Second World War. This recognised that the most senior officer on a squadron was not necessarily best qualified to lead it on some specific operation. It has been common practice for a highly experienced NCO pilot to lead a fighter section in which the other aircraft were flown by less experienced officers; or for a flying officer to lead a newly arrived flight lieutenant or squadron leader on a fighter or bombing task. On bombers, it was usual for a new and out of practice squadron commander to make a first trip as second pilot. On this mission, Captain Conran led, with Major Salmond, squadron CO, as his observer. All three aircraft scored direct hits which set light to the target: the leader making three runs at 100 feet before bombing.

Second Wing was given the railway station at Menin and the railway junction at Courtrai as targets; while Third Wing was allotted railway stations at Douai, Lille and Don. These were attacked at 3 p.m., while the ground force was preparing for its second assault. No. 5 Squadron's Captain G. I. Carmichael hit railway lines near Menin with a 100-pounder from 120 feet, was rocked by its blast and hit in the engine by a rifle bullet. Captain Strange, of six Squadron, in a BE2c, once again put up an outstanding performance. In bad visibility he was flying below the 3000-foot cloud base when flak sent him up to shelter in it, and find his way through by compass. At Courtrai he descended to rooftop level to go for the station. A sentry began shooting at him. Strange shuffled him off this mortal coil with a well-aimed hand grenade, and a decisiveness Hamlet would have envied, before dropping his three twenty-pound bombs on a stationary troop train. His aircraft had nearly forty bullet holes, but a later report revealed that he had killed or wounded seventy-five troops and disrupted traffic for three days.

Louis Strange was neither a hot-head nor merely an inordinately brave man who appeared to thrive on taking risks and was gifted with superb flying skill. One had merely to catch a lively glance from his intelligent eyes, see the animation in his handsome aquiline face and sense the humour that accompanied his quick shrewd brain, for his ability to strike one. It was he who had devised a mounting that allowed a Lewis gun to be fitted at an angle so that it fired outside the propeller disc; and, with Second Lieutenant R. B. Bourdillon, of the Intelligence staff at HQ Third Corps, had made a bomb sight "consisting of a couple of nails and a few lengths of wire", which was widely adopted until a more scientific instrument was invented.

The battle raged for two more days, on the last of which a storm blew. Reconnaissance and bombing sorties continued when weather allowed. Mapplebeck, who, with Joubert de la Ferté, had flown the RFC's two first sorties of the war, was shot down near Lille, evaded capture, reached Holland and returned to his squadron a month later. He was killed in a flying accident that August. Ludlow-Hewitt, then only a captain on No. 1 Squadron, acquitted himself less well than Strange. Flying a BE8, he bombed a railway bridge and missed; then dropped a 100-pound bomb on a railway station. His aim was true and he hit it fair and square. Unfortunately, whereas he thought he was over Don, he was actually bombing Wavrin.

In April, Lieutenant Rhodes-Moorhouse won the RFC's first Victoria Cross. Ordered to bomb an enemy concentration at Courtrai, he went down to 100 feet to drop his 100-pounder under intense fire. Mortally wounded, he found the strength to return to base and insisted on making a full report before he would allow the Medical Officer to attend to him. He died next day.

The bad weather hampered the enemy, who, even when it was fine, crossed the lines less often than the British and French, and there were no stirring single combats between rifle-armed pilots and observers. The principle of including tactical bombing of specified targets as part of a battle plan had, however, been established. But bombing would have to remain a crude and secondary task until the bombs themselves had been made more effective and available in far greater quantity, and an accurate bomb sight devised. The emphasis was still on reconnaissance: but aircrew, and Henderson and Trenchard, in whose hands their fate lay, were eager to have, and aeroplane designers to produce, machines that would out-fight the enemy one-to-one in the ultimate form of air conflict they foresaw and which was now drawing near.

The German air force was more concerned with its French counterpart than with the British. L'Aviation Militaire's numbers exceeded the RFC's; its Voisin two-seaters were equipped with a Hotchkiss machine-gun. Some had a 37-millimetre cannon. No gun approaching this calibre would be seen again until Hurricane IICs were armed with 40-mm cannon in 1941.

Lieutenant Colonel Barès's superior intellect had already perceived that strategic bombing was an essential accompaniment to tactical bombing. The First Bomber Group, consisting of three Voisin esca-drilles, formed in late 1914 under the command of Commandant Goys, had been supplemented by the Deuxième and Troisième Groupes de Bombardement, flying Voisins, Caudrons, Farmans and Bréguets. To ease the pressure on their Russian allies at the Eastern Front, the French air force attacked tactical targets in both the French and British sectors of the Western Front, as well as strategic ones in Germany.

In April, an explosives factory at Buss, a power station at Rombach, blast furnaces at Thionville, near the German frontier, and armament factories in the Ruhr had been hit. None of these raids met any opposition from flak or defending aircraft. When the Germans fired chlorine gas shells into the French lines for the first time, at Ypres on 22nd April 1915, they invited a fearful reprisal. It was delivered on 26th May: Commandant Goys led his three escadrilles off at 3 a.m. on a five-hour sortie to bomb acid and chlorine works at Ludwigshafen and Oppau. The standard French bombs were 90-millimetre and 155-millimetre shells fitted with fins. The First Bomber Group dropped eighty-three of the former and four of the latter on this operation. Everyone returned unscathed; Goys some weeks later than the rest: after a forced landing with engine trouble, he was taken prisoner but escaped.

The German civilian population vented their anger and resentment at these unopposed incursions by scrawling a distinctly uncharitable graffito on walls in the afflicted towns: "*Gott strafe England und Unsere Flieger*: God strike England and our flyers." Blaming the British was a gratifying unearned tribute to the little RFC.

The next major assault had begun soon after the Germans' use of gas at Ypres. On 9th May the French attacked on a four-mile front between Lens and Arras. The British, attacking Aubers Ridge, which lay beyond Neuve Chapelle, made use for the first time of a kite balloon, lent by the French, which they used for artillery observation.

Tactical bombing was tried again, but with less happy results than on the first occasion. Planned to damage railway lines, stations and rolling stock, HQ buildings and bridges, it failed because no direct hits were scored.

Artillery co-operation with both British and French batteries was, however, successful. Many enemy guns were knocked out, thanks to six hours' flying by Lieutenant C. B. Spence in a 16 Squadron Farman, with Second Lieutenant the Hon. W. F. Rodney as his observer: until a shrapnel shell shot down their aircraft and killed them.

Ground the Allies had won on the opening day they lost on the following. The battle then gradually waned, until it was resumed on the 15th as the Battle of Festubert. On 16th May, Nos 2 and 3 Squadrons bombed through the morning mist, and again scored no direct hits. This battle faded out by 27th May in the British sector, owing to the abundance of enemy machineguns and a lack of shells for the British artillery. It dragged on until 18th June in the French sector.

In July, Henderson ordered an analysis of British and French bombing between between 1st March and 20th June. This showed that 141 sorties had been flown against railway stations, of which only three had achieved direct hits that did the desired degree of damage. Destruction of lines was quickly made good, and accuracy over junctions was spoiled by increasing flak, and, for low-flying aircraft, machineguns.

For the time being, Henderson issued an order that, pending the training of special bombing squadrons, a proportion of pilots in all squadrons would be trained in bomb dropping. Bombing under the orders of Army Commanders would be limited to such objectives as Headquarters, telephone exchanges, munition factories and areas covered by close reconnaissance. Lines of communication bombing would be done only under orders of GHQ. He also held a conference between the British and French air Services, at which it was agreed that, for the next Allied offensive, they would together plan the bombing of lines of communication.

Henderson did not favour one form of operation at the expense of another. He was as conscious of the part that purely fighting aeroplanes would play in the future as he was of the bombers' rôle. Although air-to-air combat was still incidental to all other flying, he and most senior officers in the Allied and enemy air forces knew that the day of great air battles would come. From the first his idea had been to form squadrons exclusively composed of fighting aeroplanes. Meanwhile, the few of these that were available were divided among the squadrons, so that each had one or two. Barès had come early to the same conclusion and at the turn of the year l'Aviation Militaire already had one such escadrille allocated to each Army.

Single-seater aircraft, armed if possible with a machinegun, had come by the incongruous name of "scouts", which was more appropriate for those engaged in reconnaissance: scouting and reconnaissance being synonymous. With their love of sport and tradition of fox-hunting, it is strange that the British did not give the name "hunters" to what would later become known as "fighters". The French had introduced the equivalent term early in the war, with their escadrilles *de chasse*.

The whole question of designing an aeroplane for the specific purpose of challenging enemy aeroplanes to fight began from the basic difference between all flying machines of that period: the positioning of the propeller astern, which had given it its name, or in front, when it was properly called an airscrew, although the word "propeller" was always used. In two-seater British aircraft, whether pusher or tractor, the observer sat in front of the pilot and manned the machinegun if there were one. This meant that the tail of a pusher was unprotected. In a tractor type, the arc of fire to the rear was restricted by the bulk of the pilot; and the propeller cut off a large segment of the forward field of fire.

The pilot of a BE2c was unenviably placed. This aircraft now carried a Lewis gun as standard. A spigot protruded from its underside and there were holes around the rim of the front cockpit in which it could be located. The observer had to heft the bulky, twenty-seven-pound gun from hole to hole, according to whether he wanted to fire to the right or left forward, abeam, or at a stern quarter. That was hard work for him, swaddled in several layers of bulky garments, in danger of rupturing himself as he shifted the weapon without dropping it overboard while the light little BE2c dipped and swayed, bucked and skittered, and panting with effort in air that became noticeably thinner as an aircraft climbed above 5000 feet. For firing astern, a hinged and swinging arm was fixed abaft the front cockpit. And now, although the pilot was spared vigorous exertion, life became acutely disconcerting for him, with the hot blast and crackle of bullets whizzing past his ears; and worse when tracer ammunition was invented, and flashed, multicoloured and scintillating, past his eyes as well. It needed only some sudden jolt from turbulent air to misalign the gun and kill him.

The French crew of a rear-engined Voisin did not suffer from this inconvenience. They sat side-by-side instead of in tandem, with the gun mounted immediately behind their shared cockpit: which allowed almost an all-round field of fire.

While British and French manufacturers were still designing both rear- and front-engined types, the Germans were producing mostly the latter; and all recognised that it was in these that the future lay, not least because they were the faster. The new generation of German aeroplanes would be armed and have a greatly improved performance. For the time being, as they were tractor types, and the observer occupied the rear cockpit, their armament would consist of a rear-mounted Parabellum machinegun with a field of fire covering the tail and both beams; but unable to shoot dead ahead.

The shooting down of the French pilot Roland Garros on 18th April 1915 brought a change of fortune. It threw the research and development programmes in all three countries out of balance. It set off an acceleration in the competition to produce an aircraft that would dominate all others.

Garros combined cerebral with practical excellence. Cultured and sensitive, he was also a leading professional test pilot before the war, working for the Morane-Saulnier company, and an expert at aerobatics.

When he joined l'Aviation Militaire and was posted to a Morane single-seater escadrille, his chief concern during the early weeks of war was to devise a means of firing through the propeller disc. Much professional thought had already been given to this in many quarters. As early as 1912 the possibility of firing a gun through the propeller hub itself had been explored. Failing this, for bullets to avoid hitting the propeller blades, the engine revolutions and rate of fire of the machinegun would have to be synchronised. The ideal solution would be to use the engine to fire the gun. Some mechanisms had been patented, but none worked satisfactorily. A Swiss engineer, Franz Schneider, had invented one which he had tried on a French Nieuport and German LVG.

Monsieur Saulnier, of Morane-Saulnier, had also designed one, but the Hotchkiss ammunition was unreliable and frequently hung fire, so that a round was fractionally delayed, causing the bullet to hit a blade. Unable to overcome the problem of faulty rounds, Saulnier resorted to a cruder method, with which others elsewhere were also experimenting: fitting deflectors on the propeller. Some bullets would pass between the blades. Those that did not would be turned aside without harming the propeller. This device had not actually been tried in the air.

Towards the end of 1914, Garros obtained leave from his escadrille to return to the factory to perfect the device and to try it out in flight. He changed the shape of the blades, so that a narrower portion came in line with the gun and a smaller deflector could be used. Months of trial and error, during which he frequently shattered a propeller and had to glide to ground, finally produced the definitive article: a wedge-shaped and channelled deflector which guided rounds away to one side and was linked to the propeller shaft with strong braces. Garros also found that bullets either of copper, or lead thickly jacketed with steel, had to be used. Any other type would shatter on impact.

In March 1915 he rejoined his squadron. He was all the keener to demonstrate the efficacy of the new deflectors because General Headquarters had shown its total indifference to this great innovation by cancelling an order for several Moranes to be modified with it. On 1st April he had his chance, as described in the opening paragraphs of this book.

He had scored five kills and was on his way to bomb the railway station at Courtrai on 18th April, when he ventured dangerously low and one rifle bullet from a German soldier, Private Schlenstedt, smashed the Morane's petrol pipe. Garros had to land. He was setting fire to his machine when enemy troops seized him.

Twenty-four hours later the propeller was at the Fokker factory in Germany. Anthony Fokker was Dutch, and therefore theoretically neutral. He had put his genius at the disposal of Britain, and been rebuffed. He offered his services to the Germans, who instantly accepted. Five weeks later, Fokker presented the German air force with two of his new single-seaters, each fitted with a machinegun and not only deflectors (which he claimed to have invented himself) but also an interruptor gear that co-ordinated the gun's rate of fire with the speed at which the propeller rotated.

The single-seater unequipped with either innovation could still be lethal, as Captain Lanoe Hawker, DSO, a flight commander on No. 6 Squadron, demonstrated. Twenty-four years old at the time, a scion of the landed gentry and conspicuous for his ramrod bearing and impeccable turnout, his comrades referred to him as "jolly old Hawker" on account of his serene disposition. An exceptional shot, he is the subject of legends which must, in the absence of Combat Reports to substantiate them, be regarded as apocryphal. It is variously stated that he shot down enemy aircraft with his deer-stalking rifle, a regulation issue infantry Lee Enfield .303 or a Belgian cavalry carbine. It is also claimed that he could unerringly hit an enemy pilot in the head at remarkably long range. It is true that he flew with a rifle propped against his right leg, ready to hand, even when flying a BE2c, although the observer had a Lewis gun. One of his observers complained that he never knew when bullets from his pilot's rifle would whip unpleasantly close past his head, to supplement his own machinegun fire.

On 25th July 1915, flying a Bristol Scout with a *Lewis gun* on a mounting of his own design, which aimed the weapon at an angle of forty-five degrees to his line of flight, he scored two victories and won the RFC's second Victoria Cross: for, it is often claimed, bringing down *three* enemy aircraft with *rifle* fire during one sortie. His Combat Report – No. 93 – differs. He was on patrol at 10,000 feet when he saw two German aeroplanes. "The Bristol attacked two machines behind the lines, one at Passchendale about 6 p.m. and one over Houthulst Forest about 6.20 p.m. Both machines dived and the Bristol loosed a drum at each at about 400 yards before returning." One made a forced landing behind enemy lines, which was confirmed by No. 20 Anti-Aircraft Section.

"The Bristol climbed to 11,000 and about 7 p.m. saw a hostile machine being fired at by anti-aircraft guns at about 10,000 over Hooge. The Bristol approached down-sun and opened fire at about 100 yards range. The hostile machine burst into flames, turned upside down, and crashed E. of Zillebeke." When Hawker saw the enemy machine burst into flames and throw the observer out as it overturned, he was as upset as Garros had been. That other country gentleman and expert shot, Manfred von Richthofen, who had been an avid slayer of birds, deer and boar from boyhood, would not have shared this distress.

Hawker's VC was awarded after these two successes in recognition of all the daring patrols he had carried out over a long period.

CHAPTER 7

1915. Gird Up Now Thy Loins Like a Man

The air forces of Britain, France and Germany were maturing fast. Aviator types were emerging, recognisable by their dress, comportment and esoteric language. Among the RFC, one didn't crash, one "committed crashery". To pupil pilots, the same opprobrious term as used for the enemy was applied: "Huns". An aircraft that had been shot down was "fanned down", and the man who did it was probably a "hot stuffer"; if he flew with tremendous dash and a touch of recklessness, he was a "split-arse merchant", who had probably "bunged off" in a hurry when he left the ground. To feel afraid was to "get the breeze up" and to become excited was "getting into a flat spin". Non-flying officers were "kiwis".

The RFC alone had its own uniform: a high-necked double-breasted tunic with no buttons exposed, known as a "maternity jacket", worn with breeches, ankle boots and puttees. But this was compulsory only for other ranks. There was a wide variety of dress among the officers, who could wear riding boots, or, off duty, slacks and shoes.

Officers who entered the RFC direct from civilian life wore the official rig. Those who had transferred from another arm might do so for walking out or ceremonial occasions, but for flying preferred to wear out their old uniforms, with a pilot's or observer's brevet on the left breast and the RFC badge often replacing a regimental one. Thus there was a profusion of turnout to be seen on every squadron, even the kilt and tartan trews. The RFC headgear was a forage cap, the "fore and aft", worn at a rakish angle. Officers who sported their previous uniforms might keep to a Service Dress cap or a glengarry.

L'Aviation Militaire was more caustic than the RFC in its jargon. Generals were referred to as "fatras", "trash". "Coucou", "cuckoo", meant an aeroplane, also called a "zinc", the word commonly used for small cafés, which had a zinc counter. To crash was to be "dans le décor", and the mechanics who did their best to avert such a misfortune were "les rampants". This has an unfortunate connotation that would not be acceptable in a British Service. As an adjective, "rampant" means "servile". In one sense it is associated with the ground, but in a sneering way. "Lierre rampant" is "ground ivy" and "un style rampant" means "a pedestrian style": so reference to the latter does at least imply a pun, and one with no malice, but if the allusion was to ground ivy, and the intention was to suggest that the mechanics hugged the ground rather than risk the air, then there is a regrettable taunt implicit. The engine was a "mill", "moulin". If it went, you said "il gaze", "it goes strongly". If it didn't, the predicament was a "carafe", apparently derived from "rester en carafe", meaning "to be left out of it" or "to stop short". Then you would return to the pilots' hut, "la cagna", while it was repaired.

In dress, as could be expected, Frenchmen showed both flair and style in their multiformity of fine – even flamboyant – plumage. *La Guerre Aérienne*, a weekly magazine, had this to say of the contemporary pilot: "Black pea-jacket, khaki tunic or light blue dolman, red breeches with sky-blue stripes, black with scarlet stripes, complete uniforms of iron grey, 'horizon' or mustard, hard or soft caps, red tarbooshes [worn by some colonial regiments], berets, silk ties."

The same publication gives a description of a typical pilot which could just as well have been applied to the RFC and Luftstreitkräfte. "Flying seems to have a sort of fascination, and specially when it is combined with the attraction of combat. Perhaps this is because the game is new, but more probably on the whole because people know so little about flying.

"Thus, the air arm exercises an irresistible attraction for fighting men, because it is new and a little mysterious. Every soldier who wishes to avoid routine is therefore interested in this arm. One can, besides, see in this a desire to get away from the anonymity of the infantryman. The aviator finds a certain dignity in war: he sees it as a means of distinguishing himself from the mass of other fighting men." It quotes a military pilot: "For us all, flying constitutes a resurrection of our personality, disappeared during the months in the trenches." Another pilot's comment is: "Suppleness of muscles, power of the lungs, stamina of the heart, professional ability are nothing without the sincere, ardent desire by which one recognises the real pilot."

Displaying the innate national cynicism, or at least a realistic recognition of human nature, with reference to the above: "In consequence, the majority of pilots come from other arms. One is therefore led to question the deep motives of these volunteers. Two golden wings stylishly embroidered on the tunic collar confer in the eyes of the public an enviable distinction on those who wear them. Flying bestows, on some, glory of the purest kind, but to all its elect it assures a small worldly success about which the least that can be said is that it agreeably tickles the fibres of human vanity.

"One is thus led to distinguish between several types of volunteers: he who joins from love of flying and whose will to serve never fails; he who seeks greater public attention, admiration that flatters his vanity. In short, who thinks, right or wrong, that on a squadron he will find more shelter from danger. Indeed, after a mission, hours of liberty follow during which one forgets the dangers one has run and prepares for the morrow's work in tranquillity and safe from danger. This situation, relatively privileged, can easily be contrasted with that of the infantryman, who stays two weeks at a time in the trenches among mud and lice." Many of those who volunteered to fly, in all the combatant nations, did so simply to get away from the danger and discomfort of the trenches.

But the commentator abandons his stricture against those who seek vainglory and an avoidance of discomfort and danger by joining an air force, when admitting: "Certainly the majority of pilots are good chaps, who enjoy danger, dream of nothing but mischief and fighting, hurl themselves headlong into adventures..." Yet he reverts to the Frenchman's ingrained scepticism, adding: "But others demand of flying substantial guarantees and the certainty of glory."

Among the German aircrew who wrote with relish of their experiences, Baron Elard von Loewenstern shows up as a man who would not have been out of place in the British or French air Services. "The first necessity," he says, "is to acquire a feeling for flying. Here there opens to one the possibility of a new way of life. Flying! – to take part in laying the foundations of a new arm; also the wish to leave the monotony of military service."

There was no humbug about him. He gave quite frankly one other good reason for transferring from the ground arms. "I will be honest and say a word about the financial improvement one obtained from flying pay and a handsome daily subsistence allowance from the moment one began to fly. Entry into aviation thus offered material advantages." And he a nobleman!

The Baron was not a cynic. There was a confession of the romantic about his transfer to the Aviation Service in his acknowledgment of an attraction to flying and to being in at the birth of a new Corps. And when he reported to the flying station at Lawica/Posen, he at once responded to the unique atmosphere of an airfield. "Lined up in front of the big hangars stood the aeroplanes. A smell of oil, petrol, fabric dope and paint lay in the air. Engines thundered. All this completely gripped me at once.

"Shortly after I reported to the Station Commander, Hauptmann von Poser, an NCO looked me up and down appraisingly, and asked if I would like to take a short flight with him. He had to see how long it took a newly delivered Mars to climb to 800 metres. The invitation came rather suddenly. I walked over to the aeroplane and looked it over carefully. I had heard that the mechanics referred to it as 'Old Mrs Mars' and that there were many adverse stories about it. Could this crate really climb to 800 metres in 40 minutes? But curiosity triumphed. I put on a crash helmet and thick scarf and went aboard.

"Men were crawling about in front of, under and around the aircraft. Suddenly the propeller began to howl. It began to turn faster and faster. Everything – I with it – shook and vibrated. But the Mars itself did not move forward. The chocks were under the wheels, which I had not noticed. 'It doesn't want to go,' I thought. However, as the propeller began to scorch around, the chocks were quickly pulled away. The machine began to roll slowly over the ground. We had covered three-quarters of the field and still the good Mars was on the ground. 'Out,' I thought to myself, 'the beast doesn't want to take off.' Then, before I was aware of it, we were airborne. The wind whistled, I felt the blast."

Soon he is rhapsodising about his training as an observer. "Then began a splendid life for us. Because I was light, I was able to make many flights when various others could not, on account of their weight. Slowly we learned to behave like professionals. We learned that there were two types of weather, namely 'flyer's weather' and 'flying weather'. Flyer's weather justified idling about, when there was fog or rain. Flying weather meant we could fly. We learned expert expressions. An aeroplane was a 'crate', the propeller a 'lath', a pilot an 'Emil' and an observer a 'Franz'."

Why Franz? Because a pilot, one day, couldn't recall his observer's name and called him by this one; which is like shouting "Bill" anywhere in England or "Jock" in Scotland, whereupon several men will appear in response. By unusual extension, however, "to observe" soon became "franzen", "to Franz". The observers, in retaliation, referred to their drivers as "Emil"; although there was never a fabricated verb "emilen", "to pilot".

When he completed his training, von Loewenstern went to the Eastern Front and it was not until 1916 that he was posted to France.

The intention to form an Italian air Service was formally declared by a Royal Decree of 25th October 1914. It is interesting to recall two paragraphs in the document, which sanction the conceptions of the period.

"Reconnaissance is entrusted in the most part to the cavalry, to cyclists and to aviation. Aviation can contribute effectively, whether by long-range reconnaissance or short-range. But methods are still being developed, hence are as yet few. However, indications are that their employment will have only a general use.

"Dirigibles and aeroplanes serve essentially for strategic reconnaissance and *in exceptional cases for tactical reconnaissance*. Air reconnaissance, besides being vast, quick and comprehensive, is able, in favourable conditions, to give information about the general situation rapidly to the Commander of the ground forces."

At a distance, then, of a few months from the oubreak of war, it was not seen that a unique feature of aerial observation lay in its potential for tactical use. It was, on the contrary, precisely this function that it fulfilled, on a growing scale and with increasing reliability, throughout the four years of the war.

Like the Germans, the Italians decided that the task of observation was so important that it must be restricted to officers, and entrusted piloting, regarded as secondary, to NCOs and the rank and file.

The Italian Military Air Corps – Il Corpo Aeronautico Militare – was constituted by a decree dated 7th January 1915. The records pay this tribute to its progenitor and first Director General: "It is to the foresighted initiative, the clear vision, the vigorous firmness, the generous activity of Colonel Moris that is owed his success, against everything and everyone, in giving life to an organisation that the tragic reality of war would develop."

It was organised in two Commands: one comprising balloons and airships; the other, aeroplanes. In March 1915 a Group of artillery squadrons was formed; and, two months later, a Group of civilian flying schools for volunteer pilots.

In the first months of 1915, of all the raw materials necessary for the construction and repair of aeroplanes, Italy had only a modest quantity. The sole engine manufacturer was the Gnome Company, with twenty employees and a production of one engine a month. The firms building aeroplanes were Società Italiana Transaerea, Savoia, Nieuport-Macchi, Oneto and Caproni, each of ten to twenty workmen and all producing six to ten aircraft a year.

When Italy declared war against Austria on 24th May 1915, the Corps was mobilised and put under the orders of General Headquarters. It had 20 officer observers, 91 pilots on the squadrons, 5 flying schools, and, at the end of May, 200 pupil pilots.

Aircraft strength of the combatant countries was: France 1150; Britain 166; Italy 58; Germany 764; Austria–Hungary 96.

Francesco Baracca, a twenty-three-year-old lieutenant of cavalry, who was to become famous as Italy's most successful fighter pilot, had been sent to the Reims flying school in 1912. His instructor praised him to the Commanding Officer as particularly gifted. "He has sensibility, sharp sight, control over his nerves. I have tried several times to take him by surprise by unexpectedly cutting the engine, sudden nosedives and banking steeply. He has never been perturbed. He is undoubtedly a first-class pupil, the best of the Italian party."

Baracca was a regular, with the conventional attitudes of his breed. In contrast, Fulco Ruffo di Calabria, seventh Duke of Guardia Lombarda, who was five years his elder and ended the war as Italy's fifth-highest-scoring fighter pilot, was a wartime volunteer; and a markedly more volatile character. At the time when Baracca was learning to fly, the Duke, who was a senior executive in an Italo–Belgian company operating steamers, and trading, in Senegal, rushed home to fight a duel. He noted in his diary: "Two days in Naples to buy clothes, two in Rome to sabre my opponent, a return ticket to Brussels and Aja to embrace family and friends, then immediate return to the colony."

Two days before Italy entered the war, Baracca was one of six officers who were sent to France again, this time for instruction on the Nieuport-Macchi, with which the squadrons were equipped. On 7th June, he wrote home from Paris to say: "We have to start almost from the beginning, because the old type of Nieuport that I have been flying has completely different controls from those of any other aeroplane. The Nieuport that we will have to fly, which has a speed of 145 to 150 kilometres an hour, is difficult, and has to be handled with care because it needs a big aerodrome for take-off and landing." On 19th July he returned to Italy and No 1 Fighter Squadron.

Fulco Ruffo di Calabria, meanwhile, had come back to Italy in 1914 to raise funds there and in Belgium with which to set up his own trading company. When his country went to war, he remained and rejoined the cavalry, in which he had done his compulsory military service. Realising how little use would be made of the cavalry, he decided that fighting aboard an aeroplane would be congenial to his temperament and volunteered for the Air Corps.

He was sent to the flying schools at Turin and Pisa and had no sooner passed his wings test on 15th August 1915 than he immediately wanted to venture the most complicated aerobatics: spins, sideslips, the falling leaf, loops (at that time called "hoops of death"). But he was not yet experienced enough to execute these correctly and had several crashes. One day he landed with the wings of a Blériot warped by the stress he had imposed on them in a prolonged spin.

The flying schools did not instruct aerobatics in dual-control aircraft. The instructors' function was to transform with the maximum care a pupil who, nine times out of ten, had never seen an aeroplane close up, into a pilot capable of flying within narrow limits, namely, without damaging himself or his aircraft.

On 28th September, Ruffo di Calabria was made operational and on 1st October he joined the Fourth Artillery Squadron. The main task of the artillery squadrons was to direct battery fire and carry out visual and photographic reconnaissance. For this they flew the Caudron 63. The crew comprised a pilot and observer. It barely attained a speed of 100 kilometres an hour and was unarmed because its engine's lack of power imposed a choice between the weight of a wireless set and of a machinegun.

As the official monograph on his career puts it: "From the first day of the war Ruffo di Calabria brought to light those qualities which, more than any, enabled him to add his name to the roll of the greatest pilots of the first world conflagration: flying skill, fervid aggressive dash and total disregard of risk. He quickly began to distinguish himself. On 12th November 1915 he earned a eulogy for a reconnaissance mission carried out at an altitude of a few hundred metres under sustained enemy fire on the Lower Isonzo."

What he really wanted to fly was a fighter, but he would have to wait until the next summer before this wish was granted.

Baracca had been in combat three times already. On the first occasion he fired two or three bursts, then the gun jammed. "My machinegun," he wrote in his diary, "is a new weapon. We don't know it well and the fault is to some extent ours. It is badly placed in the aeroplane and to shoot is a very acrobatic business, and I lost faith in being able to do anything." On his second and third encounters with the enemy he suffered the same frustration. It was not until 1916 that he scored his first victory.

*　　　*　　　*

Gabriele d'Annunzio made his first operational sortie on 7th August 1915, in a Naval Air Service seaplane. Admiral Cutinelli signalled Naval HQ as follows. "1350 hrs executed offensive demonstrative action Trieste by two Italian seaplanes two French. The machine flown by Miraglia observer d'Annunzio dropped a tricolour flag on the city announced the Poet about to drop an explosive bomb on the ammunition store Maria Teresa near Sanità. On account of damage to chute no more bombs dropped. Bologna dropped four bombs, of which three seen to explode with good results same wharf. One French machine dropped four on gasometer without assessment result. Intense machinegun, rifle and cannon fire from ground, enemy aircraft took off to intercept. Despite all this all returned Venice unharmed. Machine Miraglia d'Annunzio hit machinegun bullet possibly explosive starboard stern fuselage shattered."

Following on this the Austrian Government put a price of 20,000 crowns on d'Annunzio's head.

The poet's (or Poet's, as the official record styles him) own account of the failure to release all the bombs is less drab and telegraphic. "We had carried eight bombs, to drop them on warships and fortifications around the enslaved city. The first seven fell on their intended targets. When it came to the turn of the eighth, the release mechanism malfunctioned. The bomb did not drop: it remained half-outside the aircraft, refusing to fall, despite every effort. The situation was tragic. The Austrian aeroplanes chased us, forcing us to turn for Venice.

"At every jolt of our machine the bomb could have exploded. We felt ourselves lost; but there were graver dangers! The bomb might explode over Venice. The idea oppressed me and tortured me. Never in my life had I suffered such anxiety. Anyway, while I continued to pump petrol with my left hand, with my right, outside, I hung on to the bomb with tenfold effort.

"At last we passed over the Lido and the buildings of Venice. Thanks to the incomparable skill of the pilot we landed gently in the basin, sheltered from the wind. All was saved!"

Not a bad bit of self-advertisement, one might say.

After this and other flights made by d'Annunzio in seaplanes based at Venice, Lieutenant Pilot Miraglia, as squadron commander, sent his Headquarters a report which helped to obtain for d'Annunzio an aeroplane observer's brevet.

Although the campaign on the Italo–Austrian Front is regarded as a minor one in comparison with the battles at the Western Front, it was important enough for the British, French and Germans all to send ground and air reinforcements there at various stages of the war. The physical setting and its effect on aerial operations is described in sonorous periods by Generale A. Felice Porro in his official history, *La Guerra Nell'Aria*. "The high, deep Alpine arch, with the lofty peaks of its rocky masses, constituted a monstrous obstacle to the flight of the small aeroplanes along the whole front from Adamello to Monte Nero.

"The formidable Alpine barrier seemed to wish, at the beginning of aerial hostilities, to make a solemn affirmation of its power, demonstrating that, though man had conquered the third dimension, the earth of ages towered into it with its pinnacles of rock and ice, still capable of resisting audacious human attempts to overfly them."

The aeroplanes, endowed with engines of low power, attained modest heights with exasperating slowness; their flight was precarious because of the unreliability of these engines, therefore particularly dangerous in mountainous regions where places suitable for a forced landing were few.

Edward Mannock, who was the most successful British fighter pilot of that war, and second among the Allies, with only two fewer kills than the Frenchman, Fonck, who topped the list, had not even joined the RFC in 1915, although he was in his twenty-eighth and twenty-ninth years.

His antecedents for becoming a pilot, let alone one of such eminence, or an officer, were unpromising.

Born on 24th May 1887, he was the son of a regular soldier, a drunken Irish corporal in the Royal Scots Greys and later in the 5th Dragoon Guards, whose own father, of all incongruous occupations for one with such progeny, had been editor of a Fleet Street newspaper.

At the age of ten, in India, Edward Mannock went blind in both eyes for two weeks, from a dust-borne amoebic infection. He recovered his sight, but there was corneal damage to his left eye, whose vision remained poor. His brutal sot of a father used to taunt him about his eyesight and tell him he would never be a real man, i.e. a soldier. He found that a determined show of aggression was enough to unnerve the bully. When his father threatened to beat him and advanced with raised fist, the boy did not shelter behind his mother, but stepped forward, inwardly quaking, to confront the man, who would desist. He said it proved to him that even a bogus display of fearlessness discouraged an adversary.

After the Boer War, Corporal Mannock, having served his time, was unemployed. When he had spent all that he and his wife had scraped together, he deserted her and their three children. The boy Edward had to go to work: first for a greengrocer, at two shillings and sixpence a week, then for a barber at five shillings. In 1903 he became a clerk in the National Telephone Company. Three years later he transferred to the Engineering Department as a labourer, which meant climbing telegraph poles to repair wires in all weathers. He also joined the Territorial Army, in the Royal Army Medical Corps, and rose to sergeant.

He was a ranting, proselytising Socialist, which was not surprising in view of the penurious circumstances of his youth. But he was also deeply patriotic to Britain, despite his Irish ancestry, and would verbally attack anyone who expressed anti-monarchist or anti-British sentiments.

Taken by an urge to go abroad, he sailed for Turkey on 9th February 1914, in the hope of finding work in a new cable-laying operation there. He did so and was made a supervisor.

When war broke out Germany began to negotiate an alliance with Turkey, but the work of building new telephone exchanges and laying cables went on. In November, the British Ambassador and his staff returned to England. Germans were in control and all British expatriates were declared prisoners of war. The Turks arrested them and imprisoned them in a communal cell. They intended to deport the prisoners, but the Germans objected. After Mannock and others had made several attempts to escape, the British were moved to a concentration camp.

Mannock, who had led his companions in hammering on the cell door and singing, now arranged that a Turkish visitor, who had worked for him, would cut the wire and get him out of the window every night to go and buy food for them all. He was caught and put in solitary confinement. He was as persistent a nuisance to his captors as that other great pilot and leader, Douglas Bader, some thirty years later.

Thanks to American intervention the British were repatriated in January 1915. Mannock arrived emaciated and ill with malaria, another legacy from childhood years in India.

His closest friend, Jim Eyles, a civilian and militant Socialist, described Mannock's intense hatred of the Germans. "When I told him of the most recent actions, especially the German gas attacks, his blood ran hot. Even his waxy complexion could not conceal it. His face reddened and I saw his knuckles growing white as he clenched and unclenched his fists in a growing fury." Fellow prisoners also said that he had declared the intention to "get into the Army and kill as many of the swine as possible".

In July 1915 he rejoined the RAMC and was soon a sergeant again. He thought that his comrades lacked aggressive spirit, but if they understood more about Germany and Turkey, they would acquire it. He used to harangue them about this. When demonstrating how to treat wounded, his lectures were lurid. He described front line dressing stations in gruesome detail – mud, filth, enemy bombardment, blood and mangled limbs – which whipped up his own imagination at the same time. When it occurred to him that he might have to attend to a German battle casualty, the idea so repelled him that he went to his CO and asked for a transfer to the Royal Engineers as an officer cadet. Nobody, not even himself, would have suspected, as 1915 drew to its close, that Edward Mannock, known, on account of his Irish birth, as "Mick", would go down in history as perhaps the best fighter leader of the Great War.

James McCudden, who was later to be one of Mannock's flying instructors, and to spot him as a future star performer, was still waiting to be released from his duties as a sergeant air mechanic and air observer, to go on a pilot's course, when 1915 came to its end. In July he had been home on ten days' leave, after eleven months at the Front without even one day off duty. On his return, he was employed more as an observer than as a technician and had some comments to make on some pilots that reveal as much about himself as about them.

In September 1915 C Flight of No. 3 Squadron had four officer pilots and one sergeant pilot. There were four officer observers, one corporal; and Sergeant McCudden, who recorded, after a short flight with Lieutenant Ridley: "I did not enjoy it much, for the pilot was one of the most dashing and enterprising kind. Such flying is all very fine for the pilot, but not always for the passenger."

McCudden had a different opinion of Major Ludlow-Hewitt, the squadron commander. On 13th December 1915, artillery spotting at 7000 feet, they were under constant anti-aircraft fire for two hours. "I can see the pilot now," he recorded three years later, "tapping away at his key, with a shell bursting out on a wingtip and then one just ahead, not flicking an eyelid, and not attempting to turn or avoid the numerous shells that were continually bursting. As for myself, I was in a terrible state of funk, as I could do nothing but keep a look out for enemy machines, and watch the 'Archie' bursts."

At the end of December he noted: "By now, having flown a good deal with Major Hewitt, I intensely disliked ever going up with anyone else, for I can assure you that I knew when I was flying with a safe pilot, and I had now so much faith in him that if he had said 'Come to Berlin,' I should have gone like a shot." An airman's view of what makes a safe pilot is not perhaps understood by others; Ludlow-Hewitt had subjected him to two hours of intense danger at least once.

It was also in December, in an L type Morane-Saulnier, that he had his first encounter with a Fokker firing through its propeller. It approached head-on and above. There was no gun mounting on the Morane for firing over the propeller, so he had to shoulder the Lewis gun: which demanded a fair amount of strength at that height, with the lungs short of oxygen, and at any altitude with the aircraft pitching and swaying. He fired as the Fokker flew overhead and past on the right. It came in again from astern and beneath. He told Captain Harvey-Kelly, his pilot, to turn, then shot at their attacker again. "The Fokker appeared rather surprised that we had seen him, and immediately turned off to my left rear as I was facing the tail." Next, it climbed 300 feet above and dived to fire, but pulled out at once when McCudden let fly with the Lewis. It retired to a distance of 500 yards and withdrew there each time it came close and was met with a burst.

That was the first air fight of a man who won the VC, DSO and MC in the next three years. It made his successful repelling of the Fokker all the more remarkable that its pilot was the redoubtable Max Immelmann.

Among those who would achieve less distinction, but render service that was equally valuable in its way, Algernon J. Insall deserves a place. Although he was at school in England, his home was in Paris, where his father was a doctor. On the declaration of war, Dr Insall closed his practice and returned to England with his wife and two sons.

When, with his elder brother Gilbert, Algernon Insall volunteered for the infantry on 25th August 1914, he was only seventeen. The recruiting officer told him he would have to add a year to his age. He returned next day and did so. Both young men were accepted by the Public Schools and Universities Brigade, in the 18th Battalion of the Royal Fusiliers.

Early in 1915, the RFC called for volunteers to be trained as pilots, and they applied. Their reception at Brooklands was typical of the spirit that prevailed throughout the flying world. The Adjutant greeted them, put them at their ease "in a few cheery sentences", and they left his office as probationary uncommissioned second lieutenants. They would be treated as cadets and enjoy the privileges of their rank (few enough for the lowest form of officer life in any Service) without being entitled to a uniform.

They were accommodated in a pub, the Blue Anchor, where the pupil pilots messed in a private dining room. "A pleasanter start to a new life it would be hard to imagine," Insall recorded. He and his brother were welcomed by their fellows.

Among the many reminders of sudden death to which pupil pilots have always had to accustom themselves, Insall experienced a particularly unnerving one when a senior instructor joined him for early breakfast. Affable and chatty, but pressed for time, the instructor bolted his food and hurried off to air test a Blériot. Insall, dawdling over his hard-boiled eggs, was surprised by the sudden entrance of the Station Sergeant Major, pale with shock; to announce that the Blériot had crashed and the windscreen had "sliced the top of the pilot's head off as though it were an egg". Insall, who was due for a lesson after the air test, found his appetite for boiled eggs gone, and for flying not quite as sharp as it had been.

Having qualified on pushers, he was posted to Netheravon to convert to tractor aircraft, on No. 11 Squadron, which was equipped with the Vickers Fighting Biplane, VFB5, commonly known as the FB5 Gunbus, preparatory to field service. This was the first homogeneously equipped squadron in the Service and its aircraft, although a two-seater and a pusher type, was the first purpose-designed fighter in the RFC. An instructor there, who had served at the Front and sported an enormous goatskin greatcoat of the type some French officers wore, was evidently in the same class as the instructor who used to bawl at Collishaw and other pupils in Canada. "Use your flaming rudder," he would bellow. "Your rudder, I said. Are you bloody well deaf? Go on, turn her round. *Round*, I said, not over, you bloody idiot."

Being taught by others was unalloyed bliss. Having asked "Anything I must not do, sir?" Insall was told "You go ahead and do just what you like. But for your peace of mind and mine, don't hold her nose up when you see the air speed fall below, shall we say, forty-eight miles an hour, or let it drop beyond, say, sixty. If you do anything silly or dangerous I shall probably belt you over the head."

He was doing well until, practising landings, he came in short, touched cart-wheel ruts in the grass and the aeroplane somersaulted. He hurt his knee and was grounded for ten days. When he resumed flying, he found he felt physically sick when approaching for a landing. He went up with one of the best instructors, made seven attempts to land, but found he could not overcome his inhibition. The squadron was almost ready to go to France. Insall ceased pilot training and remained with it as an observer. In the last week of July 1915 he arrived at the Western Front.

Corporal Georges Guynemer, one of the earliest French aces, was already at the Front. Serious, ascetic and religious, he had been suspected, as a boy, of being tubercular and still looked frail at the age of seventeen years and eight months when war broke out. For this reason he was twice rejected when he tried to enlist. At the third attempt, in November 1914, he was admitted as an apprentice aviation mechanic. Having achieved that important step, he pressed on until he was accepted for flying school. In April 1915 he got his wings. The next month he was posted to Escadrille MS3, which flew both one- and two-seater Morane-Saulniers. In July, with Private Guerder as air gunner, he scored his first victory and was awarded the Médaille Militaire. Four months later he had added a kill in a single-seater to his tally and was made a Chevalier of the Legion of Honour.

The air war was still a fairly casual and leisurely affair. One Sunday forenoon Guynemer shot down an enemy machine over Compiègne, where his parents lived, and was not sure where it had hit the ground. Seeing people coming away from Mass and knowing that his parents would be among them, he landed in a roadside meadow, hailed his father and asked him to search for his victim. Soon after, he brought down another aircraft in flames, near an artillery battery. He landed, as pilots often did, to have a look at his handiwork. The captain commanding the battery ordered a complimentary salvo to be fired, then tore gold braid off his képi and gave it to Corporal Guynemer "to wear when you are also a captain".

But the pace and ferocity of aerial combat were about to become much hotter.

CHAPTER 8

———◆———

1915. The Slaughter Starts

Henderson had fallen ill in March and after a spell in bed was ordered to convalesce in the south of France. He returned two months later to find that Sykes, whom he had trusted despite his unsavoury reputation, had been intriguing against him, insinuating that the RFC ought to be commanded by a younger man in the best of health; and, of course, lobbying for his own usurpation of the post of General Officer Commanding.

He also learned that Trenchard was abed with one of the severe migraines that had afflicted him since he was badly wounded in the Boer War. He went to see him, told him that he had discovered Sykes's turpitude and was dismissing him, and invited Trenchard to take on the vacant post.

Trenchard refused. He admitted that he had criticised the organisation and the policy of the Corps, and would do so again if he thought it justified; but this was not because he was scheming to replace Henderson. He suggested Brooke-Popham, whom he praised as loyal and able. Henderson concurred.

It was as well that Headquarters was rid of Sykes. A more dangerous and difficult period than any yet was about to assail both the RFC and the Aviation Militaire.

It was a paradox that as the arming of British, French and German tractor aeroplanes with forward-firing machineguns proliferated, the lethality of air fighting became greater and the kind of death that pilots and observers suffered became increasingly horrible, so the courteous generosity of spirit displayed by both sides grew more pronounced.

On 15 May 1915, Rothesay Stuart Wortley noted: "A curious kind of comradeship appears to be developing between the two opposing Flying Corps; more especially since fighting in the air has become rule rather than exception." He was prompted to this by the fact that each side dropped notes on the other's aerodromes to report when a machine had come down behind enemy lines, and the fate of its occupant(s). He added: "One would imagine that the German Air Service has attracted the pick of the German officers," because they were behaving with courtesy and consideration, even tacitly offering commiseration.

British pilots frequently dropped challenges to single combat in some appointed place and at a set time, but these were not being accepted. This contradicts his praise by implying cowardice; which could hardly be a trait found in the pick of any nation's officers. One RFC pilot, after many attempts to spur someone up to fight, contemptuously dropped a pair of old boots with the insulting note: "Herewith boots, pairs one, German flying officers on the ground for the use of."

The mutual behaviour of adversary airmen did not really amount to chivalry. The word implies the good sportsmanship and manners of a fencer who, having disarmed his opponent, steps back to allow him to pick up his weapon and fight on. It was certainly displayed on the battlefield when an eighteenth-century French general invited the British enemy to fire the first volley. Between pilots and observers who were trying to kill each other, not merely prove who flew or shot better, true chivalry would have meant sparing an enemy's life. That enemy might very well kill the man who had spared him, or someone else on his side, in a later encounter.

The consequences of the Germans' capture of Garros's invention spread quickly. Fokker's monoplane, the Eindecker, the EI, introduced large-scale slaughter into air fighting, in comparison with which the killing and wounding done with carbine or handgun seemed trivial. With it, it brought an increased danger of fire, the most dreaded cause of death.

The EI was no dazzling performer, with its top speed of 81 m.p.h. and rate of climb that took it to 3000 feet in seven minutes. It was not even original: pretty much a copy of the Morane-Saulnier Type N, which could attain 90 m.p.h. and climbed rather more quickly. It was only a few miles an hour better than the BE2c, and 10 m.p.h. faster than the Gunbus. But it was nimble, and, above all, extremely difficult to see in the sky. It had such thin wings that, head-on, they became visible only when it was probably too late to take effective evasive action. Even then, it was the round shape of the fuselage that made it identifiable, while the wings still appeared to be faint streaks across the background of clear sky or cloud.

Fokker, with Leutnant Parschau, demonstrated two of "his new" machines, which were really a rehash of others' inspirations, to Feld-fliegerabteilung 62 at Douai, on which were serving Boelcke, who was already making a name for himself, and the still unknown Max Immelmann. Fokker and Parschau fought mock combats and fired at ground targets. The spectators were impressed. GHQ had told the Dutchman to find an enemy aeroplane and give the most convincing display possible of the EI's superiority by shooting it down. He found a potential victim, but, to his credit, was too squeamish to make the kill and turned away without having violated his neutrality any more than he was already doing by aiding and abetting the Germans. He was soon virtually a prisoner in Germany, with citizenship forced on him and his life forfeit to his masters if he tried to return to Holland.

By July, eleven EIs had been delivered to the Western Front. The German Air Service made an error of judgment that was fortunate for the Allies. Instead of forming whole units completely equipped with them, it spread them around the existing Abteilungen so that soon each had two. On Flight Section 62, Boelcke was charged with doing most of the proving.

If the Fokkers had been formed into homogeneous units and used to punch with maximum force on successive parts of the Front, both the French and British air forces would have suffered even worse casualties than they did. Nevertheless, their appearance began a dire period for the Allies.

At the end of July, Lieutenant Colonel Brooke-Popham wrote to tell Lieutenant Colonel Arthur Ashmore, commanding the Administrative Wing, in England, that the German aeroplanes were becoming far more active and making a regular habit of attacking reconnaissance sorties. The RFC was having to fight for all its information. It was also having to fight in pairs as well as singly and it would probably become necessary always to send aircraft out in pairs or even whole flights. General Henderson therefore wanted the squadrons at home, which would be sending replacements to France, to practise formation flying and doing simple manoeuvres in couples.

Manufacture of the new Fokker was delayed by the unreliability of the Oberursel engine. While the Germans waited impatiently for the fighter to reach the Front in large numbers, a fairly leisurely and relaxed mode of operating survived a little longer. War in the air was always an intensely individual matter, each kill very much a personal success. In the first twelve months it not only demanded initiative and was fought aircraft-to-aircraft, which meant, for pilots, man-to-man, but it also bestowed a freedom that would not be possible when the whole process became formalised.

If the word "fighter" had not yet come into use, the terminology was at least rapidly approaching it. "Fighting" and "fighting type" aeroplanes were written and spoken about. Henderson saw that the most effective way of using the single-seater scouts that passed for fighters in the early days would be to form squadrons exclusively equipped with one such type: as the French did. His subordinates were less perceptive and insisted that they should be distributed in penny packets around the squadrons. Obviously the pilots held this view because they all wanted a chance to get their hands on one. Hence the opinion held by the wings and expressed to the General Officer Commanding contradicted his own. He should have imposed his views, but did not. Trenchard would have.

Henderson probably gave way because only the Bristol and Martinsyde Scouts with a Vickers gun on the upper wing were available and neither was the true purpose-built fighter for which he was looking. On 25th July 1915, however, No. 11 Squadron became operational at the Front, fully equipped with the Vickers Gunbus. Although this was a two-seater, the squadron was the first to have only fighters and these all of the same type. Nos 5 and 6 Squadrons each had a Gunbus flight.

In July, the Fokker E2, which had a 100 h.p. Oberursel engine in place of the E1's 80 h.p., began coming into service. This put its speed up to 87 m.p.h. and gave it a 12,000 foot ceiling. But some of the engine problems lingered.

Reconnaissance sorties were increasingly threatened. On 21st July, Pilot Corporal V. Judge, in a No. 4 Squadron Voisin, and his observer, Second Lieutenant J. Parker, armed with a Lewis gun, a rifle and two revolvers, were outnumbered and brought down on enemy ground by machinegun fire. One of the enemy flew over No. 4's aerodrome and dropped a message to say that the observer had succumbed to his wounds; and "the German pilots have the highest praise for their opponent, who died in an honourable fight".

Ten days later, Captain J. A. Liddell of 7 Squadron won a VC, flying an RE5 with Second Lieutenant R. H. Peck. An E1 dived on them. Bullets hit Liddell's thigh. He fainted. The RE5 lost 3000 feet and when its pilot recovered consciousness he found it was on its back. He rolled out straight and level, with a smashed control wheel and throttle. Still under fire and at low level, he managed, after some thirty minutes, to land behind Allied lines. He died of his wounds.

*   *   *

The two German pilots who most conspicuously took advantage of their new fighter were Boelcke and Immelmann. They were both thorough and methodical, treated air fighting as a science, systematically worked out its basic principles and arrived at the best ways of applying them. Their theory, which was again followed in the Second World War and remains basically valid today, was as follows.

Opening an attack, a fighter pilot should have a height advantage and the sun behind him.

He should make use of cloud to conceal his approach.

His objective should be to get so close to his target that he could not miss it, yet be in a position where the enemy could not bring a gun to bear on him. In a fight between two forward-firing single-seaters, this meant being astern of, and slightly above, the other aircraft. A single-seater fighting a two-seater that had a rear machinegun needed to position itself behind and slightly below.

If attacked from ahead, a pilot should turn directly towards his adversary. This presented the smallest target and reduced the time in which the other man could aim and shoot.

If attacked from behind, a pilot should bank into the tightest possible turn. This made it difficult for the enemy to get on his tail; and, if he turned inside the enemy, he had a chance to get on *his* tail.

Boelcke, with his greater number of flying hours and more mature character than Immelmann, was the world's first great fighter tactician. Immelmann's outstanding contribution to aerial combat was the turn which is named after him. Aerobatics are, of course, the basis of attack and evasive action.

Before the war, only a few pilots had dared to try such movements, but a surprising variety existed and some were possible only with the very light, slow aeroplanes of the period. Adolphe Pégoud, the famous French stunt artiste, demonstrated his range at Brooklands in September 1913. One item was the bunt or outside loop, a manoeuvre still forbidden in the RAF. Pégoud added two elaborations to it. In one, he half-rolled at the bottom of the bunt, so that he was upright and flying in the direction opposite to that in which he had entered the outside loop. In the other, he described a vertical S, by making half an outside loop, then pulling the stick back and making an equal semi-circle with an opposite convexity. Another amazing feat was to stall at the top of a 45-degree climb, control his aeroplane in so masterly a fashion that it slid backwards in a parabola, then, when it again momentarily hung stalled and motionless, this time as though suspended by its tail on a sky hook, he put on power and dived into his next trick.

Immelmann designed his turn to make a second attack on a target in the shortest possible time after the first. Having dived, he pulled up into a loop, at the top of which, when, of course, he was inverted, he half-rolled out so that he was upright. He would now be above his target, behind it, but flying in the opposite direction from it. By stalling and falling away into a stalled turn, he was able to dive straight back onto the enemy. There were, therefore, two distinct parts to the manoeuvre. First came the "roll off the top", which might in itself be all that he needed to do if he were taking evasive action rather than breaking off to renew an attack. Next came the stall turn, which took his enemy by surprise.

Immelmann, Boelcke and an observer, Teubern, found a small empty house in Douai and moved in. One can imagine the long discussions which occupied their evenings: the two pilots arguing, the observer making tentative contributions. None of the three smoked, they almost never drank, all were fond of sweet cakes. Immelmann wrote home to say admiringly that Boelcke was an accomplished sportsman "in the most varied directions". He described him as "an extraordinarily quiet, reflective fellow and he owes it to his mode of life that he shows no sign of the strain, although he has been flying from the beginning of the war and has spent a long time at the Front".

In another letter to his mother, on 25th June, he tells her that he has been flying 25 to 30 kilometres behind the enemy lines and has made the most flights on the Abteilung: 21, compared with Boelcke's 19.

When Fokker demonstrated the E1 to Flight Section 62, he was impressed by Immelmann's ability. He also admired him because he neither drank nor smoked and went to bed at ten o'clock. Learning that this paragon of rectitude and seemly hours was an engineer, Fokker offered him a job after the war, as pilots with his qualities would be hard to find. Fokker even went into detail. He would start as chief test pilot and progress to an engineer's job when he had acquired the necessary knowledge of aeronautical engineering. He would be paid a salary plus a percentage on the sale of machines and the fees paid by pupils whom he trained. Immelmann was not as callow as he looked. He assured his mother that this offer had been made in front of witnesses.

On 1st August, the British bombed Douai aerodrome. Immelmann records waking at 4.45 a.m. to hear about twenty aeroplanes overhead, bombs bursting, flak firing. By then, 62 had the Fokker EII. He and Boelcke took off. He says he saw both British and French machines, although there were no Frenchmen there. His aircraft recognition seems to have been poor, even making allowances for the pale light of dawn, for the BE2c was unmistakable by anyone with average eyesight. Both British and French aeroplanes bore roundels, but the RFC's consisted of a blue outer circle, a white inner circle and a solid red "bull's eye". The Aviation Militaire had the red and blue reversed.

Immelmann attacked a BE2c that was being flown without an observer – to allow an increased bombload – hit it and forced it down. He had mistaken its markings, so, landing alongside, he called in French, "You are my prisoner." The pilot held up his arm and replied in English: "My arm is broken. You shot very well." Immelmann then repeated himself in English. He helped the pilot to climb out of the cockpit and to remove his leather coat and tunic. The arm was badly wounded. Immelmann made his victim comfortable on the grass and sent for a doctor. This was not an example of that much misused word "chivalry": it was kindness.

He had already been awarded the Friedrich Augustus Silver Medal for gallantry in the face of the enemy, after several artillery spotting and escort missions. He was now awarded the Iron Cross, First Class, for attacking a superior force and bringing one enemy down. Judged by later standards of what had to be done to win a medal, both these seem to be what, in the Second World War, the RAF called "a piece of cake".

Boelcke, who had also won an Iron Cross, and Immelmann kept pace with one another until by mid-September each had three victories. On 11th October Immelmann went ahead of his rival by shooting down a BE2c.

Political intrigue in London, coupled with a feud between French and Kitchener and Churchill's quarrel with the First Sea Lord, had led to upheavals and changes in Government. Lloyd George was made Minister of Munitions. Henderson had to decide what would be the best way to ensure the greatest possible strength of the Military Aeronautics Directorate in fighting its corner against the other Services.

To stand up to the powerful protagonists for Land and Sea at the War Office and Admiralty, whoever led the RFC at the centre of power and influence would have to be a general who had earned high repute for his technical knowledge and experience of battle. He had to recognise that he was the only one who met these criteria. Accordingly, with typical self-abnegation, for he would far rather have stayed in the field and would perhaps have risen higher than lieutenant general, he took the War Office appointment upon himself and designated Trenchard as his successor.

On 19th August 1915, Trenchard was gazetted GOC RFC.

Henderson found that, to meet the demands which came to him from every theatre of war, he needed to the full all his great qualities. He had often to fight for his Corps in circumstances where there was no tradition to support his argument and the rôle of the new arm was not fully understood. To the end he remained unruffled and kindly in judgment of those who did not understand, but he alone knew what a strain this imposed on him. About one vital factor he was never disturbed, for he knew that in its new chief in France, the Flying Corps had someone whose personality must impress itself in the difficult days ahead on a Service highly responsive to the inspiration of its leaders.

One of those most affected by changes was Maurice Baring. He had been sent to Italy to visit various factories and to see the Caproni bomber, find out if Beardmore engines would be suitable for it, and if it was advisable to order a Caproni for the RFC.

When he returned to RFC HQ he found Henderson gone. Trenchard told him he would give him a month's trial and asked if he would like to stay on. Baring asked permission to go to London to consult Henderson. Trenchard agreed. Henderson told him he would do best to stay in France. His occupation thenceforth was to trail round with Trenchard and take notes of everything from the General's demand for Oxford marmalade for his breakfast to spares needed by the squadrons.

By all accounts he did an impeccable job and won universal affection, respect and esteem. He began as a square peg in a round hole and ended as a perfect fit. He was obviously an immensely likeable man and utterly unassuming despite his acclaim as a writer.

A period of change in air operations had dawned with the arrival of several Fokker Eindecker at the Front that coincided with the next big ground offensive. On 25th September the Battle of Loos had begun. The French launched the main attack in Champagne, while a combined French and British attack was made in Artois. Both were badly planned. Joffre made a fool of himself. The result was a futile carnage of the kind that happened repeatedly in the Great War. Insignificant gains of ground were made at a cost of 242,000 Allied casualties. The Germans lost 141,000. Haig had been against the plan. After the battle, he replaced French as Commander-in-Chief.

The RFC had recently begun escorting reconnaissance flights by at least one machine, a fighting type if possible. Often it was only one BE2c accompanying another.

The Aviation Militaire had, since the summer, equipped its fighter escadrilles with the delightful little Nieuport 11, known as the Bébé, capable of 90 m.p.h. and 15,000 feet. Despite its performance, it was at a disadvantage against the Fokker because it was armed with a forty-seven-round drum-fed Lewis gun mounted on the upper mainplane. It would take more than the Nieuport's margin of superiority in speed and climb to compensate for the Fokker's Spandau fed by a 500-round belt and firing through the propeller. Oddly enough, and seldom mentioned, the Bébé had first seen service with the RNAS, at the Dardanelles.

Coincidental with the Battle of Loos, which ended on 13th October, the Germans woke up to the wisdom of forming homogeneous single-seater units and set up three. Both Boelcke and Immelmann were chosen to join one. While Immelmann stayed at Douai, Boelcke was sent to Rethel, facing the French sector of the Line, to take part in an innovation: barrage patrols, singleton standing patrols near the Allied Lines, waiting to pick off enemy aircraft with the Fokker's now established dive out of the sun.

The French bomber groups, GB 1, 2 and 4, were busy with daylight raids during the offensives in Champagne and Artois. On 2nd October sixty-two aeroplanes attacked Vouziers, the site of German HQ. Two were shot down. After this, Nieuports escorted them (shades of the German air fleets of 1939 and 1940 over France and England, with their protecting Messerschmitts), but this of course reduced their range. After a raid by nineteen bombers escorted by three fighters on 12th October, the bomber groups began to practise formation and night flying. Captain Happe, who commanded MF29, introduced large formations and took his escadrille out in the V that was to remain standard in all air forces for decades. He did not lead, but gave high cover, searching for the enemy and warding off attacks. After his men had bombed a poison gas plant at Dornach on 26th August and the Aviatik works at Freiburg, the Germans put a price of 25,000 marks on him. He sent them a message to say that, to avoid wasting time on anyone else, they should note that his aeroplane was recognisable by its red wheels.

Despite the alarm that the Fokker created as soon as it came on the scene, at the end of October there were only fifty-five at the Western Front, and, two months later, eighty-six. As significant as the E1's advent was the small number of Albatros D1s that began to appear on the scene: the first aeroplane fitted with twin machineguns firing through the propeller.

On 7th November, Lieutenant Gilbert S. M. Insall, of No. 11 Squadron, brother of the Algernon J., an observer on the same squadron, who had been put off his breakfast boiled egg at Brooklands, took a Gunbus up with Air Mechanic T. H. Donald in the observer's seat. They saw an Aviatik and chased it. The German lured him towards a rocket battery, but he divined the enemy's intention and was able to position himself so that Donald emptied a drum into its engine. Both machines descended below cloud. Donald fired a second drum and the Aviatik thumped down on a ploughed field. Pilot and observer tumbled out and tried to turn their Parabellum on the Gunbus. Insall dived, his observer fired again and wounded one of the Germans; who now began to run for cover, one supporting the other. Next, Insall dropped an incendiary bomb on the enemy machine which was at once hidden by smoke. On the way back to base, flying very low, ground machinegun fire hit the Gunbus in its fuel tank. Insall had to land behind a copse 500 yards behind the French lines. Enemy artillery fired 150 shells at the Gunbus, but none hit it. At night, under screened lights, the crew repaired the damage and, at first light, they flew home. This earned Insall a VC and Donald the Distinguished Conduct Medal.

The younger Insall had something to say about the realities of air escort during the Battle of Loos. B Flight of 11 Squadron, commanded by Captain Playfair, was detached to Brancker's wing to provide a protective screen over the battle area. "Four 60–70 m.p.h. rumbling old aeroplanes [Gunbus], each with a stripped Lewis gun guaranteed to blue its barrel if more than ten rounds were fired per burst, two .455 automatics or revolvers and a stripped Short Lee Enfield were responsible for the air safety of some thirty-odd observation and reconnaissance aeroplanes and thousands of troops defenceless against air attack." They spent eighteen days on detachment, maintaining almost continuous patrols over the First Army battle area at 8000–9000 feet. They saw the Germans use gas but never an enemy aircraft close enough to engage it.

On returning to their squadron, at Vert Galand, they made their first landing in the dark by the light of flares. Landing grounds were lit by setting out petrol tins, their tops cut off, containing half a gallon of petrol and cotton waste. These were laid out and lit on all RFC aerodromes at the Front when any aircraft was overdue and extinguished when all had been accounted for. The flares were arranged in an L shape and the pilot landed towards the short stroke.

The RFC was still disgruntled by the reluctance of the Fokker pilots to accept challenges. The press had been writing about Immelmann and it was known that he was in the British sector. Early in November 11 Squadron's CO obtained permission to invite Immelmann to meet one of his pilots in single combat. Playfair was chosen. Insall volunteered to be his observer-gunner, as they usually flew together. The rendezvous was to be a point over the trenches between 10 and 11 a.m. on any day from 15th to 30th November. The message was never dropped, but, astonishingly, when an RFC aircraft was shot down during that time, the crew were asked the name of the officer who had been selected to fight Immelmann. German Intelligence must have been excellent.

Soon after Christmas, Immelmann forced down Lieutenant Darley of 11 Squadron behind the German lines. Darley's thumb was almost severed. Immelmann landed beside him, took out a knife, briskly completed the amputation, bandaged Darley's hand and saw him off to hospital and prison camp.

In contrast with the French bomber groups, the RFC's bombing was still, in many ways, rough and ready. Insall recalled forced-landing with Playfair at an aerodrome where a "Horace" Farman, a combination of the Henry and Maurice Farmans, was being prepared for a raid. Mechanics were tying bombs made of howitzer shells, fitted with vanes, to the undercarriage; with binder twine. Over the target, the observer would have to climb out of the nacelle and cut the string.

Insall also gave an instance of the extreme discomfort of flying in cramped, unheated aircraft. After a three-hour flight at 7500 feet, one December day, he and the pilot were unable to move for three full minutes. When they did climb out, they fell, unable to straighten their legs, and mechanics had to help them to their feet and support them to the rest hut.

Immelmann went on leave only once, in order not to miss any chance of action. On 28th November he took part in a flying display at Leipzig to raise funds for Christmas presents for all members of the Air Service. At a formal lunch, the mayor presented him with a silver cup and praised him as a most important member of the young air arm.

Of the many clues to his character that his habits betrayed, one that might indicate either extreme patriotism or capricious eccentricity was manifested at Christmas, when he sent his mother 100 marks and his sister 50, to buy themselves presents, as he did not want to spend the money in France.

Trenchard, with the lives of every member of the RFC at the Front in his hands, believed that air supremacy must be gained and retained at any cost; and that if he reduced his effort, the enemy would concentrate strong formations on various parts of the British area.

There was good co-operation between the British and French air Services. L'Aviation Militaire had suffered badly against the Fokkers during Joffre's attack in Champagne. Commandant de Peuty, Colonel Barès's senior representative in the field, willingly gave Trenchard the benefit of his experience and advice. They decided that the Corps squadrons, which operated in co-operation with the Army, must be protected by a strategic offensive that would seek to destroy the enemy's aircraft deep behind his lines. Unlike Trenchard, however, de Peuty preferred to wait until his Service was adequately equipped with Nieuport 11s.

The RFC lost fifty pilots and observers in the concluding two months of 1915.

CHAPTER 9

———◆———

1916. The Very Dead of Winter

The old year went out in blistering style with a fight typifying all the constituents of aerial warfare at that time, in which three fighter pilots destined for great fame took part. In three-dimensional combat, pilots have to make the best tactical use of space while their faculties are constantly disoriented by changes of attitude and direction. Simultaneously, they have to make life-or-death decisions in fractions of a second *and act on them.*

On 29th December 1915, Sholto Douglas, flying a BE2c with Lieutenant Child as observer, escorted by Lieutenant Glen, also in a BE2c, set out on reconnaissance. Behind the enemy lines at 6500 feet they saw two Fokkers; then, soon after, four more. The Fokkers dived into an attack. With no means of communication between the aeroplanes, except for manual signals and the firing of Vérey lights, these attacks could not be co-ordinated. The slow BE2cs were able to turn inside the fast single-seaters. The British observers had a chance to fire telling bursts, even though they had to change drums after every forty-seven rounds, and more than ten in one burst could damage the barrel of an air-cooled Lewis gun. First one Fokker went down, and then another.

Glen's machine followed them. The four surviving Fokkers kept up their attacks on Douglas and Child for forty-five minutes. It was Boelcke who was the most pressing. He thought he had fatally wounded Child and ought now to despatch Douglas (whose identity, of course, he did not know) easily. He forced him down to 3000 feet. But he had been in two earlier fights on this sortie and his ammunition was exhausted. Douglas had his troubles, too. His petrol tank was almost empty and oil was running from a bullet hole in the sump.

Boelcke stayed, determined to prevent the BE2c from crossing back over the Allied line. The two pilots kept turning tightly; but fruitlessly. Douglas was on the inside, but Child was not shooting. Boelcke thought he had killed him. But it was not a mortal wound that forced him to cease fire: his pilot's violent evasive action had made him airsick. He had vomited over Douglas and was left limp, dazed, incapable. Boelcke could not turn inside his opponent, and, even if he had, his gun was useless.

At this point Immelmann came haring in, shooting, to polish off his opponent. With two Germans now after his life, Immelmann's bullets cracking past him, and with no means of knowing that Boelcke had emptied his ammunition belt, Douglas went into a steep corkscrew dive. At 300 feet Immelmann's gun jammed and the Germans broke off. Douglas held his downward spiral until he was ten feet off the ground. He crossed the lines in the French sector and put his aircraft down with a dry fuel tank and sump.

Boelcke, writing to his parents, described Douglas as "a tough chap who defended himself vigorously".

The new year made a stormy entrance. High wind and heavy rain kept aircraft grounded on most days. But on the British Front the Army was insistent on its need for long reconnaissance. Every time there was a break in the weather, the Flying Corps tried to provide it. The Fokkers were waiting. Few of the aircraft that took off returned. Nor were they merely forced down and the crews captured: the Fokkers shot them down and killed their occupants.

Unfulfilled requests for long reconnaissance, prevented by the weather, accumulated. Drastic action had to be taken. On 14th January 1916, RFC HQ issued an order which, in its clumsy ungrammatical way, was to the point. "Until the Royal Flying Corps are in possession of a machine as good as or better than the German Fokker it seems that a change in the tactics employed becomes necessary. It is hoped very shortly to obtain a machine which will be able to successfully engage the Fokkers at present in use by the Germans. In the meantime, it must be laid down as a hard and fast rule that a machine proceeding on reconnaissance must be escorted by at least three other fighting machines. These machines must fly in close formation and a reconnaissance should not be continued if any of the machines become detached. This should apply to both short and long reconnaissance. Aeroplanes proceeding on photographic duty any considerable distance east of the line should be similarly escorted. From recent experience it seems that the Germans are now employing their aeroplanes in groups of three or four, and these numbers are frequently encountered by our aeroplanes. Flying in close formation must be practised by all pilots." One asks oneself why, when "must" is used four times, this unequivocally emphatic and imperative word is weakened by "should" in two places: thus making it possible legitimately to disobey the order.

These instructions meant that, while the tasks demanded of the RFC remained as abundant as before, the number of visual and photographic reconnaissances that could be carried out was much reduced, because more aircraft had to be allocated to each. For example, when GHQ ordered a reconnaissance of Belgian railways on 7th February, the BE2c of No. 12 Squadron detailed to do it was escorted by three BE2cs, two FE2bs and a Bristol Scout, also from No. 12, plus two FE2bs and four BE8s from two other squadrons.

The casualties that the Fokker had wrought and was continuing to inflict resulted in histrionics verging on hysteria among some politicians. None of these had the sense to see that what he said in Parliament, and how he said it, would be reported in the press and could have an adverse effect on the RFC's morale. "The Fokker Scourge" was one submissive term born at Westminster, and another, applied to British airmen and thus even more demoralising, was "Fokker fodder".

On the German side, Boelcke had returned to Douai before the end of the old year and was one of the very few who habitually ventured over the Allied lines. To be fair, this was discouraged: because the Fokker's engine was unreliable and if it was not forced down, it might have to land involuntarily. The enemy would then be able to learn all about its synchronising gear. The EIII was by now in service, different from the EII only in having a couple of feet more wingspan, which slightly increased its speed.

On 12th January Boelcke scored his eighth victory. Immelmann equalled this the next day. Both were awarded the Pour le Mérite, a decoration that had never before been awarded for flying. They both received immense publicity and adulation. Immelmann, because he was the younger and looked innocent and immature, an impression emphasised rather than negated by his thin straggly moustache, attracted the greater attention. Girls sent him rosaries, crucifixes and holy medals. The letters he received reached fifty a day and he had to tell his batman to acknowledge them. The batman soon complained of a sore finger.

A momentous new factor, which would have perhaps the most far-reaching effects of any tactical decision taken by the Germans during the war, loomed over the Western Front. Field Marshal von Falkenhayn, Chief of the German General Staff, making an appreciation of the state of the war at the end of 1915, stated that England (meaning, as usual, "Britain") was the more important member of the Franco–British alliance. "The history of the English wars against the Netherlands, Spain, France and Napoleon is being repeated. Germany can expect no mercy from this enemy, so long as he still retains the slightest hope of achieving his object." The imputation to Great Britain of intent to annihilate Germany was as insolent as it was false. Germany was the aggressor and Britain had stepped in merely as a defender.

The British sector of the Front was not easily penetrable by a sustained and ferocious offensive. "In view of our feelings for our arch-enemy in the war that is certainly distressing." After some more fulmination Falkenhayn continued: "France has almost arrived at the end of her military effort. If her people can be made to understand clearly that in a military sense they have nothing more to hope for, breaking point would be reached ..."

He concluded that a massive penetration of the French front was not essential. Germany should subject France to such relentless attrition that it would, to quote Liddell Hart, "bleed to death". To achieve this was simply a matter of choosing the place on which to focus the onslaught: somewhere "for the retention of which the French command," explained Falkenhayn, "would be compelled to throw in every man they have". The choice lay between Belfort and Verdun. Verdun was chosen because it threatened German communications, because it could be isolated in a narrow salient, which would cramp its defence; and because it was the gateway through which Attila had led his Huns to attempt the conquest of Gaul in the fifth century.

Germany started preparing for the many battles and long siege of Verdun, which would transform air operations in several ways.

At Verdun experience in the use of fighters accelerated. It was here that the French and German air forces learned lessons which caused them to revise the organisation and tactical use of fighters, thus setting a pattern that the Germans adopted from the French and the British adapted.

The Germans opened the Battle of Verdun on 21st February 1916 with an artillery bombardment on a fifteen-mile front. They used almost their entire air force on the barrage patrols that they had initiated in October. The initial purpose of maintaining these standing patrols over the front line was to drive the French aircraft away. Its basic flaw was in being purely defensive. Only offensive action can dominate any air space. This is true today and was just as true then. For the time being this tactic had considerable effect, because so many escadrilles were still flying slow, poorly armed Moranes, Voisins and Caudrons. It could not bar the way to them all, though.

Barrage patrols constituted another mistake as well. By concentrating so many aircraft on them, the Germans had none to spare for what should have continued to be routine jobs: artillery spotting and reconnaissance. The only other task for which they did use a few was in close support of the ground forces.

The long-drawn-out Battle of Verdun was really a siege punctuated by furious bombardments and infantry assaults. In a siege, it is long-range cannon that are the most valuable weapons to both attacker and defender. With nobody flying artillery observation for the Germans, their big guns were being given target indications only by their captive balloons five miles behind their trench lines, whose view was much more restricted than from an aeroplane. It was counter-battery fire that was most needed and the most difficult to direct without aeroplanes. The French were getting some artillery co-operation machines across the enemy lines, but expensively. They needed to shoot down the balloons. These were defended by Fokkers, supplementing specially positioned flak and heavy machineguns. Their own batteries were handicapped by paucity of artillery observation sorties that managed to penetrate the German barrage patrols. They, too, used balloons, but these were vulnerable to attack by the enemy, despite anti-aircraft protection.

General Pétain, in command of the French forces at Verdun, saw that it was essential to take air superiority away from the enemy. Lieutenant Colonel Barès went to Verdun to organise the necessary arrangements. He increased the establishment in that sector from four escadrilles to sixteen of which six, instead of one, were to be fighters. He put Commandant Tricornot de Rose, at present Chef d'Aviation of the Second Army, in command of the latter. De Rose had the robust personality and appearance of a traditional heavy cavalryman, not least of which was his walrus moustache. He had transferred from the dragoons as far back as 1910 and obtained the first military pilot's certificate in the French Army. He now formulated the basic doctrine for offensive formation flying when he ordained that fighters must always go out in threes or sixes. Enough Nieuport 11s were available to increase the escadrilles' establishment, which was raised to twelve aircraft; an equally important measure.

Before the Nieuport Bébé entered service, some of the Morane escadrilles had made names for themselves and their commanders because the best single-seater pilots had been posted to them. The first of these was MS 3, under Capitaine Félix Brocard, a fierce-looking man of medium height with a big moustache and a body shaped like a barrel, whose habitual straddle-legged stance and straight look were the quintessence of fighter pilot aggression. The others were MS 12, commanded by Capitaine Tricornot de Rose, and MS 23, Capitaine de Vergette.

Brocard's escadrille had by now acquired Nieuports and had accordingly become N 3. Having already achieved distinction in their Morane days, the pilots had a flying stork silhouetted on each side of their aircraft, to let everyone, friend and enemy, know who they were, and were therefore called "Les Cigognes". They had been serving with the French Sixth Army and were now transferred to Verdun. The Storks, under Brocard, expanded into a group comprising also N 26, N 23, N 73, N 103 and N 167.

The Fokker EIII was not without its troubles. All three marks suffered from occasional defects in the synchronising mechanism, which led to many pilots, among them both Boelcke and Immelmann, and Fokker himself, shattering their own propellers more than once. But it was still a deadly machine to fight. Even more formidable was the EIV, which was powered by a 160 h.p. Oberursel motor that gave it 110 m.p.h. It had twin Spandau machineguns. Only a few EIVs had been manufactured as yet, but Boelcke and Immelmann each had one. It was now that the frequency of their kills began to mount rapidly. Immelmann even experimented with three guns, but found the extra weight made the EIV too sluggish.

The Fokker was not the only formidable opponent that the French had to face. Germany, after having trailed behind France in aeroplane design and development for so long, suddenly confronted her with two new single-seaters. The Halberstadt D1 had a 100 h.p. Mercedes engine and a speed of 85 m.p.h., the Pfalz had the same performance as the Fokker EIII. There were two new two-seaters, the Roland CIII, with its top speed of 103 m.p.h. and the Rumpler CI, which could attain 105.

De Rose had not yet worked out any tactics for fighting in formation, or issued instructions that this should be attempted; so the French pilots broke formation on meeting the enemy and fought individually. Given the nature of fighter pilots at any time and in those early days in particular, and taking into account the French temperament and the strong individualists that Navarre, Nungesser, Guynemer and others were, nothing else could be expected.

Boelcke and Immelmann continued to patrol singly, while the rest in the German flying units set off *and fought* in twos and threes. Boelcke, indeed, who found escorting reconnaissance sorties stultifying, was allowed to remove himself from Douai to a forward airstrip seven miles behind the Front, with another pilot and sixteen ground crew. It was a time, he said, of "Alles ganz auf eigene Faust ... Everyone on his own fist." Trenchard would have approved of the sentiment, which expressed the epitome of the offensive spirit.

The first Frenchman to gain distinction at Verdun was Jean Navarre. He was just nineteen when war broke out and he joined the Air Service. His twin brother, Pierre, went into the infantry. The first time Jean Navarre met an enemy aeroplane, he showed the stuff of which his character was made. The German flew alongside and waved. It was a foolish thing to do to a youngster like Navarre: who waved back, then put a rifle to his shoulder and shot at him. Navarre had to take both hands off the controls to do so and his Farman almost stalled, while the German dived away. Deciding that the Farman was a useless platform from which to fire any weapon, he transferred to MS 12 and by April 1915 had made two kills; and was often to be seen stunting over the trenches.

At the beginning of the Verdun siege he joined N 67. On 26th February he took off at dawn, found three unescorted two-seater enemy aircraft and promptly attacked. Two of the Germans fled at once. In the rearmost one, the observer stood and raised his hands in surrender. Navarre escorted it to a French aerodrome. Later the same morning he ran into nine German aeroplanes, picked out one, had a fight with it and shot it down.

As flamboyant as Nungesser, he had a skull and crossbones and red stripes painted on his Nieuport. Nungesser went further. His was decorated with a skull and crossbones, a coffin, two lighted candles and a black heart. After a British pilot fired at him, he lost confidence in the official markings and had an additional V in red, white and blue painted on his upper wing.

Navarre's favourite tactic was to approach his victim from astern and slightly below, then *stand up* to aim his Lewis gun. This was extremely foolhardy. The unstable Nieuport could easily have tipped him out. Navarre continued his aerobatics over the French lines and the infantry knew him as "*la sentinelle de Verdun*".

Another characteristic that Navarre shared with Nungesser – and many other French, British and German fighter pilots throughout the war – was that they both held their fire until they were very close to the target. This is habitually described as "point-blank range". It is nothing of the sort. The simplest definition of point blank is the point at which a bullet or shell begins to drop below a straight line between it and the target. Taking into account the speeds and relative positions of pursuer and target, lateral displacement and difference of altitude, plus the effect on a bullet of the wind generated by the aircraft from which it was fired, distance to point blank from the pursuer's gun could be as much as 600 yards. Nobody fired from such long range. What those who carelessly use the term "point blank" mean is a range of 25 to 50 yards. Point blank *could* fall somewhere within those limits for a revolver or pistol, but by no means for a machinegun or rifle.

Navarre's score soon reached seven, when Nungesser had six. On 4th April he shot down three German machines in the course of four patrols, but two of these fell behind enemy lines and could not be confirmed, so he was credited with only one. On 17th June he was leading a patrol of three Nieuports which intercepted three two-seater reconnaissance Rolands and shot them all down. After that, when at 12,000 feet, the Frenchmen spotted another enemy two-seater at 9000 feet and went for it. To draw the enemy's fire, so that his companions could shoot it down, Navarre swung to one side. The German observer put a bullet through his arm, breaking the bone, and then wounded him in the side. He fainted and before being able to make a crash landing he bled so profusely that he was delirious in hospital for several days and was found to have suffered brain damage from loss of blood. One glass of wine was now enough to intoxicate him. He was withdrawn from operational flying, with a total of twelve confirmed victories.

When his beloved twin, who had transferred from the infantry to the Air Service, was killed, he broke down completely. He did not return to active service until September 1918, and never flew again on operations. After the war he became chief pilot at Morane-Saulnier and had recovered his nerve enough to declare that he would fly through the Arc de Triomphe on the day of the victory parade on Bastille Day, 1919. He did not live to do it. He was practising aerobatics four days before his attempt when his engine cut at low level and his machine dived into the ground.

Nungesser, at the same time as Navarre, was also proving highly destructive to the enemy and incurring severe injuries himself. Lady Caroline Lamb's diary entry on meeting Byron was equally appropriate to him: "Mad, bad and dangerous to know." Nungesser habitually endangered his own life as carelessly as he did his enemies'. He was totally regardless of pain. His reckless style was exemplified in a cheerful understatement: "Before firing my gun, I shut my eyes. When I reopen them, sometimes the Boche is going down, sometimes I am in hospital." After having had a bad crash in 1915, he spent four months of intensive fighting at Verdun in 1916. In January 1916 he crashed on an air test. He broke both legs; and the control column smashed into his mouth, penetrated his palate and dislocated his jaw. That cost him two months in hospital. When he returned to the Front he had to use crutches, but despite this he got himself into the cockpit and went out looking for a fight. At intervals he had to return to hospital for treatment, and at the same time was acquiring more wounds and injuries. A bullet split his lip open. Crash-landing in no-man's-land, he dislocated a knee. In another crash landing with his aircraft shot up, it overturned and broke his jaw. By the end of the war he had been injured seventeen times.

At Verdun, the four fighter pilots who excelled all others were Boelcke, Immelmann, Navarre and Nungesser. They were all caught in the same current of passion for their work and swept along by it. In the two Frenchmen, delight in killing the invader predominated. In the two Germans it was absorption in the exquisiteness of their craftsmanship. All possessed an incandescent spirit compounded of dauntless mettle, superabundant aggressiveness and determination to excel.

France assembled her best pilots in virtually segregated units. The British did not approve of this system. In the RFC a squadron's successes would earn it the admiring adjective "crack". Such distinction was sought by teamwork and was won by the outstanding quality that its pilots evinced in the course of acquiring this admiration. In l'Aviation Militaire, crack escadrilles were created by posting already outstanding performers to them. The Stork escadrilles had been formed in this way, with a nucleus of leading pre-war pilots. Another that bore the same lustre was N 77, "Les Sportifs", comprising brilliant sportsmen and rich playboys; some of whom were both. One of its members captained France at rugby football, others were international racing drivers, fencers, horsemen.

Among Les Cigognes were the rich, the poor and the middling. Some counted on the money that various commercial firms put up as bonuses. Michelin, who manufactured tyres for aeroplanes as well as Service motor vehicles, paid a bonus for every victory. Guynemer totted up 15,000 francs, which, although his means were modest, he gave to a fund for wounded airmen.

Les Sportifs lived high, wide and handsome. Oozing wealth, they moved their cars, valets, mess waiters and cooks, their expensive cutlery, crockery and linen from one aerodrome to another like a maharaja's caravan. Their wives and mistresses accompanied them and were installed in the best local hotels.

Both Cigognes and Sportifs were the pets of high society. Invitations to every kind of lavish entertainment was showered on them. Cars would be sent to fetch them as far as Paris for dinners, theatres and dances when the day's flying was done.

A third colourful agglomeration was formed by the Americans who had joined l'Aviation Militaire. We have already heard of Bert Hall flying for Bulgaria against Turkey; and of Raoul Lufbery who had entered the Service as Pourpe's mechanic. They were both among this unorthodox galaxy of pilots of varied talent, united by lust for adventure and love of freedom.

Norman Prince, an American private pilot, had gone to France soon after the war began, to form a volunteer American squadron. He met another American, Edmund Gros, a doctor who had formed the American Field Ambulance Service. They set up a committee and appointed a Monsieur de Sillac as President. Dr Gros and five other Americans, one of whom was the millionaire William K. Vanderbilt, who provided the finance, served on it. Prince was not a member: his purpose was to join the unit and fly. They sought recruits among all the Americans who had already joined the French Army.

Initially, the French opposed the plan. As the static warfare sank into a morass of dreary winter inaction the notion of volunteer fighters from abroad, especially from a country so vast and rich as the USA, began to look appealing. It would be wonderful propaganda. The prospect looked all the glossier because the air forces had already acquired a romantic, individualistic image. From dissuasion, the French turned to encouragement.

The conditions the committee offered the American volunteers were generous, to compensate for their basic pay, which would be that of l'Aviation Militaire and low in comparison with American standards. William Thaw was made a lieutenant and the others would be sergeants when qualified but were joining as mere corporals. They would be given a new uniform every three months; 125 francs per head per month would be paid into the mess fund; for each confirmed victory there would be a bonus of 1000 francs. A month after the squadron came into being other perquisites were added: 1500 francs for a Légion d'Honneur, 1000 francs for a Médaille Militaire, 500 francs for a Croix de Guerre and 200 francs for each citation (a palm) added to it.

The unit was formed on 16th April 1916, under a French Commanding Officer, Capitaine Georges Thénault, and second-in-command, Lieutenant de Laage de Meux. The seven American pilots were widely assorted: one or two were comfortably off, another was a medical student, there was a Harvard graduate; there were footloose adventurers. The squadron was based at Luxeuil. Its symbol, painted large on each side of the fuselage, was the head of a Red Indian in a chief's eagle-feathered war bonnet, his mouth open as he yelled a warcry. Gros, as a doctor and head of an ambulance unit, remained a non-combatant. Prince was joined by James McConnell, Bert Hall, Elliot Cowdin, Victor Chapman, William Thaw and Kiffin Rockwell. Hall was already in l'Aviation

Militaire and had forced down a two-seater Halberstadt. They all had to go through a flying course. Their aircraft were Nieuport 11s and the unit's number was N 124. It was publicised as l'Escadrille Américaine, to which the Germans soon objected through diplomatic channels, because America was neutral. Displaying subtlety and style, the French then suggested the name Escadrille Lafayette: in memory of the Marquis de Lafayette, who had taken a group of Frenchmen to America in 1777 to fight with the colonials for independence from British rule.

The Lafayette pilots lived much like the Sportifs. Officer and NCO pilots messed together in the Grand Hotel. Their sleeping quarters were in a large private house. They did not make an auspicious start, wrecking several machines in bad landings and collisions with ground obstacles. The reporters who flocked to Luxeuil drew no veils over their bad flying or off-duty antics. Some public resentment began to grow against these pampered and apparently useless foreigners. Dr Gros had been busy finding more members for the escadrille, Lufbery among them. Necessity now demanded their presence at Verdun.

Their aerodrome on the Verdun Front was at Bar-le-Duc, where l'Escadrille Lafayette suffered the casualties that are the lot of any raw squadron. On 24th May, Thaw was the first to be wounded: in a fight with three Fokkers, when a bullet severed his pectoral artery and he almost bled to death. Bullets hit Rockwell's windscreen and its fragments lacerated his face, nearly blinding him. The next day, his head in bandages, he was on patrol again. Four Fokkers jumped on Chapman out of cloud and a bullet creased his scalp and grazed his skull. Bleeding copiously, he barely managed to return to base. On 18th June, Thénault, Rockwell, Prince and Clyde Balsley, a newcomer on his first operational flight, were attacked by fourteen Fokkers. The enemy circled them, turning inwards to fire in turns. Thénault took his men homewards in a steep dive, but Balsley could not extricate himself. An explosive bullet hit him in the stomach and wounded him severely. Surgeons removed more than twenty fragments and he lived. Chapman and Prince, flying to visit him, were bounced by six Fokkers. Chapman was shot down in flames and became the first American airman to be killed in action.

When, soon after, the Lafayettes were taken out of the line and returned to Luxeuil, Thaw had been given the the Légion d'Honneur, Chapman, Rockwell and Hall the Médaille Militaire and Croix de Guerre with one palm. They found two Royal Naval Air Service squadrons, among whom were several Canadian pilots, on the airfield. Immediate mutual liking and good fellowship were struck between the nationalities: British, Commonwealth and American. This was a feature of all relationships between the British and both American and French

Air Services. When the American Air Corps arrived in France in 1918, however, its association with l'Aviation Militaire was not always cordial.

The concentration of enemy fighters on Verdun did not noticeably afford any relief to the RFC. During the first six months of 1916 it lost an average of two aircrew a day. The loss of fifty in November and December 1915 had been severe enough. Most squadrons were still a mixed bag of two, three, or even four types of aircraft. Bewildered though pilots may have been by the daily variation of their duties between visual and photographic reconnaissance, escorting others who were thus engaged, or offensive patrols, morale remained high on most squadrons.

Trenchard wasted many lives because he equated aggression with the distance his aircraft penetrated behind enemy lines. As justice has not only to be done but to be *seen* to be done, so the RFC's aggressive spirit had to be made obvious. In Trenchard's scale of values, a patrol that went ten miles into Hunland was ten times as aggressive as one that went one mile, regardless of the quality of the work that was done when the patrol reached its limit. The man who went one mile deep might have a better chance of shooting down enemy aeroplanes than the man who went ten, but that did not seem to matter. This attitude was not only unintelligent but also cruel. British aircraft were inferior to the enemy's and even an exceptional pilot in a BE2c, Martinsyde, Bristol Scout, FE2, Sopwith, RE7 or Morane, all of which figured on British squadrons, was at a disadvantage against the Fokker, Pfalz, Halberstadt or some of the two-seaters, in the hands of a pilot who might be no better than average.

Trenchard used to visit all his squadrons frequently in his Rolls-Royce staff car. He told the aircrews that he did not ask them to do anything that he would not do himself. But he did not actually do it. This assertion was commonplace in all the Services, but its credibility varied greatly. A platoon, company or battalion commander, a flight or squadron commander, the captain of a naval vessel and an admiral at sea with his fleet, spoke nothing but the truth when he said it.

The RFC was on the threshold of better days, with new fighters soon to appear at the Western Front. Louis Strange had recently been sent back to England for a rest, and his friend Lanoe Hawker's turn had come when, on 28th September 1915, he was put in command of the newly formed No. 24 Squadron, at Hounslow. He had not had to face the Fokkers at the height of their supremacy, but he had fought for almost a year and his VC and DSO were evidence of the severe stress he must have suffered. He showed physical signs of extreme fatigue.

In January 1916 the squadron was delighted to be told that it was about to receive the new de Havilland DH2 single-seater, which had been specifically designed as a fighter. It is often said that No. 24 was the RFC's first fighter squadron and the first to be equipped with only one aircraft type. That distinction belongs to No. 11. Twenty-four, however, was the first homogeneous *single-seater* fighter squadron. And there is no accuracy in any claim that only a single-seater can properly be described as a fighter. The two-seater Bristol Fighter, when it came along in 1917, was a true fighter of outstanding accomplishments.

The DH2 was a pusher with a Lewis gun mounted in the nose. It did not look as modern as a Fokker or a Nieuport: the pilot sat in an enclosed nacelle, but behind him was an open framework of long spars, braced by struts, attaching it and the wings to the tail unit; and it had the unreliable Gnome single-valve 100-h.p. rotary engine. At sea level its top speed was 93 m.p.h., but at the heights at which it would do its work this fell to around 77. Its ceiling was 14,000 feet and to reach 10,000 feet required 25 minutes. Speeds, rates of climb and ceilings for aeroplanes were still imprecise. Much depended on how they were rigged and the condition of individual engines. Two of the same kind could have a disparity of five to ten per cent in speed and rate of climb.

Another new fighter squadron, however, beat Hawker's in the race to get to the Front. No. 20, commanded by Major G. J. Malcolm and equipped with the FE2b (familiarly the "Fee") two-seater, arrived there on 23rd January. Like the DH2, the FE2 had been designed by Geoffrey de Havilland and was a pusher with an enclosed nacelle and a latticework fuselage of beams, spars and bracing wires ending in the empennage. Following the standard British practice, the observer sat in the front cockpit, provided with a forward-firing Lewis gun. Most Fees also had a second Lewis on a telescopic mounting behind his seat, which fired upwards over the wing. The FE2 was heavier than the Fokker, so its 120-h.p. Beardmore engine did not give it quite the latter's speed, but it was equally manoeuvrable. Every pusher posed the same danger to its pilot in the event of a crash: the heavy engine was hurled forward and crushed him. On the other hand, it was an effective shield against bullets fired from astern. Of course, if the bullets stopped the engine, a resulting crash landing would flatten the pilot anyway.

Of the twelve pilots – all officers – who joined Hawker on 24 Squadron, only the three flight commanders and two others had flown on operations. Because of engine unreliability, familiarisation flying was restricted to two hours so that there would be reasonable certainty of the whole squadron arriving at its destination in France without any forced landings. Hawker was ordered to cross the Channel by steamer on 2nd

February 1916 and the rest flew over on the 7th to St Omer. They were immediately put on daily patrols at 14,000 feet from 8 a.m. until 3 p.m. Two aircraft were kept back at three minutes' notice to scramble in protection of GHQ if the enemy put in a bombing raid.

One of the marvels of all that aircrews accomplished in that war was that they did so much in such adverse physical conditions. It was intensely cold and now that patrols at heights of 14,000 feet and higher were becoming standard the low temperature was painful and incapacitating. Pilots and observers came dangerously close to suffering hypothermia. Frostbite was a common affliction. More dangerous and a greater physical handicap was lack of oxygen. In the 1939–45 war, when all aircraft were fitted with oxygen "bottles", it was compulsory in the RAF to switch on at 10,000 feet. Tolerance of oxygen deprivation varies and some men switched on sooner. Oxygen starvation causes hallucinations – clouds, for instance, are mistaken for other aeroplanes, mountain ranges, and stranger things – headache, nausea, lack of strength and energy. On top of these handicaps was the nauseating effect of castor oil as an engine lubricant. The fumes caused vomiting and diarrhoea. A tractor rotary engine sent oil spraying back over pilot and observer, reducing vision through windscreen – not all machines had them – and goggles, and blackening faces. Men had to fly with mouth and chin wrapped in a thick scarf. It was Hawker who designed, and had made, the first pair of fleece-lined thigh boots that became known as "fug-boots". These were a great comfort, but soft-soled and cumbersome and impractical for walking more than a few yards to and from one's aircraft. Aircrew who made forced landings and either had to evade capture or walk miles to a friendly unit cursed them.

Another grievous handicap in air fighting was the continued unreliability of machineguns. All were prone to jam from many causes. Fusee springs, pawls, buffers, triggers, defects in drums, imperfect rounds, all caused the frustrations of interrupted combat and lost victories.

Within a week, two of 24's pilots were killed when their aircraft spun in. The DH2 already had a reputation for involuntary and irrecoverable spinning. It was being called "the spinning incinerator", but there was nothing new about this. The Shorthorn was "the flying incinerator" and various other types were described as incinerators or coffins, because they stalled or spun easily and usually caught fire when they crashed. They were facile epithets, often intended to excuse pilots' errors. These were not always their fault, but the result of poor and hasty training that sent them into the air before they were fully competent.

The two fatal crashes were potential morale destroyers. Like everyone else, Hawker had carefully avoided a spin. He now took a DH2 up to

8000 feet and spun it several times: to left and right, with and without engine. Nobody was watching. He landed, went into the mess and announced what he had done. Everyone wanted to know how he had done it. When he had explained how a DH2 could be made to recover from a spin, all his pilots hurried into the air to practise it.

In preparation for their first fight Hawker made his pilots practise gunnery day after day, diving at a full-scale outline of a Fokker on the ground. He designed the ring sight that was adopted throughout the Service. Some of his pilots mounted twin Lewis guns and he encouraged that, not only because it would double their fire power but also because if one jammed they would have a spare. He showed the twin mounting to his Brigade Commander, who promptly forbade it. Had he had to fight in the air himself, the brigadier general would perhaps have approved of it. Hawker experimented with a double ammunition drum, one welded on top of another in the Armament Section, which led to the production of the ninety-four-round drum that was introduced soon after.

The ineptitude of those who designed gun mountings and the senior officers who would condone no modification made by the men who actually had to fight with these weapons was staggering. The DH2's Lewis was mounted on a universal joint on the left-hand side of the aeroplane's nose. The majority of pilots, being right-handed, found this awkward. Many had their guns moved to the right. It was some time before anyone did the obvious and had it mounted centrally. Another irritant was that the gun had to be held steady when being fired. This meant that, wherever it was mounted, a hand had to be taken off the joystick or throttle. In addition, the freely moving mount allowed the gun to dance and wander all over the place, even when the handgrip on the butt was firmly held. Hawker had his squadron's guns fixed rigidly; but not for long: the brigadier wouldn't allow that either. Hawker then compromised by having the muzzle anchored by a strong spring, which could not be described as a fixed mounting but did reduce most of the straying off aim.

On 24th April 1916 came the chance at last to evaluate the DH2 against the Fokker, when four of 24 Squadron escorted five BE2cs of No. 15 on reconnaissance. Twelve Fokkers attacked, some circling to prevent the BE2cs from retreating, the others waiting to make their familiar dive and zoom. When it was obvious that the reconnaissance machines had no intention of turning back, and the whole British formation was deep in enemy territory, the circling Fokkers joined their companions. Then the whole lot swarmed down in attack. The DH2s turned with an agility that surprised the Germans and made straight for them. The Germans, disconcerted by this bold tactic, pulled out of their dives and broke to right and left. In a few seconds the DH2s were into them, turning with them or inside them, firing every time they had an enemy in their sights. Two Fokkers pulled out, damaged, and a moment later a third followed. The remainder drew off and circled, trying to draw the DH2s away so that they could attack the BE2cs. The escort would not be drawn. Their job was to stick close to their flock.

The Fokkers did not attack again. . The nine British aeroplanes returned home unharmed. The DH2 had broken the Fokkers' grip on the British Sector of the Western Front.

CHAPTER 10

———◆———

1916. A Hard Life

The air forces of all the combatant nations kept their raw tenuous grip on a life that was tough, edgy and fraught with nearly as much danger from the tools of their trade as from enemy action. A raucous ironical black humour was also common property. The name which airmen bestowed on the control column, "joystick", with its phallic connotation, typified their robust brand of wit.

One innovation had already been introduced a few months earlier by the French, earned a frown of disapproval from the British and was looked at askance, but with their usual humbug, envy and imitativeness, by the Germans.

It is now necessary to take a short step backwards in time to one which, it will be seen, fits quite logically here. The great Adolphe Pégoud, whose astoundingly bold aerobatics have been mentioned, was famous also for two other feats of conspicuous bravery. In 1913 he had made the first parachute jump from an aeroplane: a single-seater in which he had gone up for the purpose, and perforce had to allow to crash after he had quit it. In the same year, he was the first man to fly inverted. He had prepared for this by having an aeroplane turned upside down while he sat in the cockpit, then suspended from its hangar roof. When he had timed how long he could sustain this rush of blood to the head, he knew the safe duration of a first inverted flight.

He is mentioned here because he was the first pilot to be dubbed an "ace". It was the French newspapers, whose reporters and readers were avidly interested in all aspects of aviation and wildly excited by the apparently romantic exploits of their military airmen, which coined the term; or title, as it began to be regarded. When, in 1915, Pégoud became the first to shoot down five enemy aeroplanes, the national press, acclaiming his achievement, sought a term of distinction for him and those who would emulate him. Ace was the chosen word.

Pégoud, sadly, did not survive to enjoy his notoriety for long. On 31st August 1915 an infantry battalion sent an urgent request for an aircraft to drive off an enemy two-seater that was harassing it. Pégoud, when not flying, was permanently standing by for a report of enemy aircraft, so that he could take off in his Nieuport Bébé at once to intercept. This time he knew where the enemy was loitering. When the German pilot and observer spotted him, they began to climb. He caught up and opened fire. The enemy replied. Presently Pégoud had to break off to reload his Lewis gun, then attacked again. The German observer put a Parabellum bullet through his heart. The Nieuport crashed to earth from 10,000 feet. His escadrille found his unlikely good luck charm, a child's cuddly toy in the shape of a penguin, among the wreckage.

L'Aviation Militaire did not at first officially recognise the status of "ace", but tacitly had to accept the word's popular use and its implication. With the passing of time it insinuated itself into official approval. Commandant de Rose saw the value of it in maintaining morale. When he took command at Verdun he officially published pilots' individual scores. This exaltation was harmless enough, good for the morale of single-seater pilots and for the reputation of the Service. The invidious effect of an elevation of fighter pilots above bomber, reconnaissance and artillery co-operation pilots and observers was apparently not taken into account. This was also the first acknowledgment of public enthralment by the glamorous notion of two knightly figures jousting chivalrously for honour and glory. The myth of chivalry has already been demolished. Nor can there be anything remotely romantic about men being riddled with bullets or incinerated alive. Honour and glory, however, remain valid.

While the Germans did not admit having adopted the ace system, they acknowledged an equivalent by setting a target of eight kills as the requirement for award of the Blue Max, the Pour le Mérite. This was later increased to sixteen. Most fighter pilots also received a silver Ehrenbecher, a beaker of honour, to commemorate their first kill. It was a brutish, ill-proportioned trophy.

To the British, conferring this sort of adulation on an individual was anathema and would have embarrassed the recipients. Air fighting, like any other military or naval engagement, was essentially a matter of team work. Commanding Officers did not like having brilliant individualists on their squadrons. As the air war developed and fighters fought in pairs, sections of three, flights of four or six and squadrons of twelve, the vindication of this became increasingly evident. None the less, and the French and Germans would indict the British for allegedly typical hypocrisy, on 6th May 1918, Lieutenant Colonel Joubert de la Ferté, Commanding 14th Wing, decreed that the minimum score for an immediate award of the newly introduced Distinguished Flying Cross, which replaced the Military Cross as a decoration for flying, would be six victories. There is no evidence of any Commander having laid down a set number of bomber or reconnaissance sorties for a member of aircrew to fly in order to qualify for this decoration. The truth remains that it is impossible to assess merit in action by any such arbitrary yardstick: five, six, eight or sixteen kills, they are all equally spurious as a real measure of what a fighter pilot deserves. They also ignore the existence of other pilots and their crews.

Despite the RFC's insistence that victory in the air was achieved by team work, individually gifted fighter pilots were more than tolerated throughout the war, as will be seen, and granted many privileges: among them, choice of aeroplane, freedom to take off when, and to roam where, they chose. And, naturally, the British press and public lauded them.

By this time the popular image of a fighter pilot was well established. *La Guerre Aérienne* really let itself go on the subject. The aviator, it said, was a man exceptional for his physical and moral qualities, an adventurer out of the ordinary; a sort of champion towards whom popular fervour was directed. This was why everyone who entered the aviation Service aspired to fly a fighter.

"Fighter pilots are an élite, a glorious élite, universally praised, officially very much appreciated. To become one of them is to receive a mark of distinction, it is the consecration of exceptional qualities."

The number of volunteers for fighters greatly exceeded the requirement. A fighter pilot was "entrusted with a supple, highly-strung, prodigiously fast machine. This seduced sportsmen by its lively charms, in particular the intoxicating speed that increased tenfold the sensation of power."

La Guerre Aérienne frequently insists on the moral qualities demanded: "What justified the fighter pilots' liberty was the fact that they put it to good use. For that, intrepidity, courage, love of sport and a taste for risk are necessary."

"My dream," wrote Brindejonc des Moulinais, "is to shoot down a dozen Boches, receive a scar and return to rest from the fatigues of summer at Val, keeping bees."

There was praise for the "daring, even temerity" of the RFC: "Every day they accomplish exploits which prove how useful sport is as a training for those who make war. Perhaps one could even reproach them for their sporting spirit, which makes them ignore danger. They hurl themselves at the enemy impetuously, with admirable bravery."

The magazine classified French recruits in four categories. First of all the cavalryman, because static warfare had grievously shrunk the uses of the mounted arm and inaction became unbearable. The infantryman: with emphasis on the need to ensure that a candidate was not seeking only to get away from the trenches. Pre-war civilian pilots formed the third group. The fourth consisted of men who had been wounded in action and were unfit to continue in their own arm. Presumably an infantryman who could no longer march or a calvary- or artilleryman who could no longer ride, but could walk to an aeroplane and sit in it, were whom the writer had in mind.

In 1914 the majority of volunteers were pre-war pilots. After 1915, "it is a man from the infantry or cavalry, who wants to escape from anonymity or inaction. Other motives are more varied: the interest to be found in a new type of risk, a taste for adventure, the attraction of a popular arm whose first appearance excited the public, sometimes also the hope of better conditions of life when not in action.

"Thus were born three categories of airmen: the real war pilot; the 'honest' pilot; the 'honorary' pilot, he whom one finds at the Front, where he flies when he has no choice: his comrades tolerate him and do not bear a grievance about his slackness, because he is a jolly fellow, a musician, an artist or runs the mess excellently."

A recruit pilot's life was no sybaritic sinecure. There were long delays before admission to flying school. Brindejonc des Moulinais, a pre-war civil pilot, complained: "We are sixteen in one room, at my request my mechanic is with me; the surroundings are better than in the foot-sloggers, but even so there are some shocking things ... my only resort is to take photographs of conditions in barracks. It is no fun and it is monotonous."

René Dorme, later of Les Cigognes, who had transferred from the artillery, was sent to the Military Academy at Saint-Cyr, where, he wrote: "They don't know what to do with us. There are thirty of us NCOs who have nothing to occupy us. I would like to shout my hatred and my disgust for those who let us stagnate like this."

Once a pilot joined a fighting unit his existence was transformed. Pilot Sergeant James McConnell, of the Escadrille Lafayette, found that during training a pupil pilot was subjected to rules and a discipline as strict as in barracks, "but once called to the firing line, he is treated on the same footing as an officer, whatever his rank. His two mechanics are under his orders. There are neither rollcalls nor other military restraints, and in place of a straw-filled mattress he has a real bed in his own room."

Brindejonc des Moulinais greatly relished the contrast between life as a trainee pilot and as an operational one. Being a Frenchman, his first observation in a letter home is about the food. "Apart from the guns which one clearly hears rumbling, this is the veritable country life as lived in a château and, my word, I should enjoy myself greatly here. Yesterday I ate wild duck. Today it will be partridge for lunch and probably pheasant for dinner. I got up at half past eight this morning and I have just breakfasted on a large bowl of milk, two fried eggs and breast of duck." In another letter he says: "Thanks to the aviator's legendary flair for sniffing out the best on campaign, the escadrille almost always establishes itself near a little village of welcoming houses and with shops abundantly provided with a variety of provisions."

Again: "In an escadrille, a pilot flys for two or three hours every day. In addition there is gunnery practice, testing machineguns, discussions and conferences. But, however, there often remain hours of leisure to spend in shooting, fishing, football, etc." And, about the men in the trenches: "I know that they get used to their life in a filthy hole, but what admiration we should have for them."

Girardot, another pilot, wrote to his parents: "I don't want you to worry about me, for it really is not worthwhile. I am very happy and I spend my life at the Front in the most agreeable way possible. As for the dangers that I run, they are small, I assure you, above all beside those of the poor rank and file, the real ones who are in the trenches."

Despite uneasy consciences, the single-seater pilots protested against any measure that tended to withdraw privileges from them and reduce them to the level of those in other combatant arms. Jacques Mortane, editor of *La Guerre Aérienne* for example, complained on their behalf about a proposal to double the number of victories needed to earn a mention in despatches from five to ten. This measure was intended to avoid arousing jealousy among the ground troops. "Our troops are more tolerant. Thank God!" Mortane wrote. "Nothing is more natural than that they should be given the benefit of praise in despatches. But it should be understood that there is a distinction between those who fight together and airmen who fight alone."

The RFC did not live as well as their French counterparts. To begin with, although the officers usually contrived to supplement the rations by local purchase, the food for all ranks was mediocre and monotonous. Occasionally officers lived in a requisitioned château or several small village houses, but usually on camp. When pasture or arable land was commandeered for use as an aerodrome, the farmhouse was probably included. It might be used as an officers' mess, and the CO's quarters or squadron Orderly Room. Almost invariably, sleeping accommodation was in bell tents and wooden huts or corrugated iron Nissens. Sometimes all ranks were under canvas, at others the officers slept in huts. There might be any number from two to six sharing a room. As the war progressed, huts for everyone became more usual.

In general, an RFC officers' mess at the Front was shabby, with poor-quality, often ramshackle, furniture. On the walls were pieces of aeroplanes the squadron had shot down and pictures of scantily clad girls. A piano was indispensable, always hard used and poorly tuned. So was a gramophone, and anyone going home on leave was expected to bring back a record or two, preferably of the latest musical comedy and music hall hits. In most messes pilots frequently let off steam by getting mildly drunk, singing, and smashing tables and chairs. The rough and tumble games that have for decades been a feature of dining-in and guest nights in peacetime were invented at the Western Front. The French and Germans frequently drank heavily but did not go in for these often bone-breaking and concussing indoor sports. Nor did the Italians or Americans.

A typical Flying Corps song, with its wry acceptance of aircraft crashes and death, was sung to the tune of "A Tall Stalwart Lancer Lay Dying". Its chorus went: "Take the cylinder out of my kidneys, / The connecting rod out of my brain, my brain, / From out of my arse take the camshaft / And assemble the engine again."

Merriment, natural to high-spirited young adventurers, was the keynote of mess life; forced though it often was, and holding a desperate note of trying to forget the next sortie. Lord Chesterfield would have disapproved. In 1749 he warned his son against, and disparaged, "horse-play, romping, frequent and loud fits of laughter, jokes, waggery ..." Between 1914 and 1918, and again between 1939 and 1945, ribald and rowdy behaviour was often what saved many men from nervous breakdowns and madness.

On some aerodromes, shellholes were lined with tarpaulins, filled with water and used as swimming pools. If there happened to be a river or canal handy, one could be sure that all British ranks would go bathing in it. By late 1916, the officers on a few aerodromes had laid out tennis courts. Often, horses were available to ride: borrowed from the cavalry, artillery or Service Corps, or commandeered from farms and large private houses. There were always several dogs on any RFC camp: some owned by individual officers and men, others the general property of the squadron.

On many squadrons, at various periods, the three flights messed separately. When two or three squadrons were brought together as a wing, they retained their own messes, separate from Wing Headquarters'. The atmosphere of a squadron in the various air forces depended to some extent on numbers and very much on the Commanding Officer's nature. Too small a mess could be claustrophobic and breed feuds. Too big a one was in danger of being cold and impersonal. The decision that an RFC squadron should consist of twelve pilots had been based on doubts that officers of high enough calibre to command a squadron could be found in suffficient numbers if the units were smaller and therefore more numerous. It was also the minimum number to ensure an agreeable variety of company. Although an escadrille and a Feldfliegerabteilung functioned similarly to a squadron, they were more closely equivalent to a flight. As has been seen, the time was at hand when several of them would be grouped together for greater efficiency.

In the British Services junior officers had traditionally been treated unkindly by their seniors: severity, contempt, rudeness, all were practised on them. In the Navy, a seventeen-year-old midshipman was liable to be given six strokes with a dirk scabbard by the Sub-Lieutenant of the Gunroom, where officers below lieutenant's rank messed.

Typically, officers, except for a handful who had been educated at home by tutors, or, like Trenchard, at boarding crammers', had been to public schools. They therefore took it for granted that initially they would be harshly received and have to serve their time before being in a position to tyrannise others and gain privileges for themselves.

Generally, in the RFC, snubs and tyranny were almost unknown: the character of the type of man who was attracted to flying ensured that. Informality and friendliness came naturally. There were exceptions. Duncan Grinnell-Milne, who had transferred from the infantry in July 1915, was given a cool reception when he reported to No. 16 Squadron in France during the Battle of Loos. He was the only new arrival that day. The aerodrome itself looked dangerous and unwelcoming: L-shaped, its perpendicular arm was 100 yards long, varied in width from 40 to 100 feet, and followed the curve of the River Lys. The horizontal arm was 20 yards long. He found a brick farmhouse containing the offices, and lines of tents. There were few officers about and the two or three to whom he spoke were off-hand and taciturn. One did ask which flight he was on, pointed, and said: "There's your flight commander."

The latter, Captain C. Wigram, asked Grinnell-Milne how many flying hours he had.

"Thirty-three, sir."

Wigram said that it was disgraceful to send pilots out with less than fifty, and preferably a hundred, hours. He asked what types the newcomer had flown.

"Longhorn, Shorthorn, Caudron, BE2."

Wigram wanted to know if it was the latest BE2 with the 90-h.p. RAF (Royal Aircraft Factory) engine.

Grinnell-Milne told him it was not.

That, said his flight commander, was no good. There were only Shorthorns on his flight and he had too many pilots already. "You'll have to wait your turn to fly. *And stand to attention when I'm speaking to you.*"

This flight and one other slept and messed aboard a barge on a canal that branched from the river. The third flight occupied the farmhouse half a mile away. When he had stowed his kit, Grinnell-Milne took a stroll. Pilots and observers whom he passed stared and did not offer a word. The atmosphere at dinner was oppressive. Only the two flight commanders conversed. The rest occasionally whispered. One sullen pilot, who had an MC for shooting down an enemy aeroplane, scowled and addressed only the senior flight commander, who responded with an occasional patronising smile.

Halfway through dinner the Commanding Officer, Major H. C. T. Dowding, nicknamed "Stuffy", for his reserved humourless manner, arrived. Wigram introduced the new arrival, who had to rise and walk to the top of the table to shake the Major's hand. Dowding tried to break the prevailing silence by speaking to each officer, but the response was never more than "Yes, sir" or "No, sir".

After dinner, when someone put a record on the gramophone, there was a shout of "Stop that bloody row."

Next evening, dinner was rather less moribund. The Battle of Loos was in full swing and the squadron had done good work. The Major sent the servants out of the room and read the communiqués and Intelligence summary. There was a secret document with a map, which laid down the RFC's duties. This had to be read and initialled by each officer.

The next morning, Grinnell-Milne was sent up in a Shorthorn with an observer. They flew to the battle front. The aircraft began to be rocked by exploding anti-aircraft shells. The observer suggested turning back. On landing, the flight commander said, "You had your first look at the lines today, didn't you?"

Grinnell-Milne could hardly credit that somebody had actually spoken to him. This emboldened him to mention that he had been "Archied".

Wigram commented that he must have been damned scared.

Grinnell-Milne admitted that he had been very scared, but that once he got used to it, it didn't seem so terrible after all.

Wigram, scathingly addressing him as "young fellow", told him not to talk nonsense and to wait until he had a bit more experience before talking lightly about Archie.

The commander of the other flight sharing the barge put his oar in: "Who says he doesn't mind Archie?" he demanded. In his turn he ridiculed the novice and accused him of thinking he knew a lot because he had "seen a couple of shells in the distance". The first time Grinnell-Milne really got fired at, he assured him, he would "go all goosey and run like a hare".

There was some discussion about a pilot who had been shot down on reconnaissance. The same flight commander was indignant because the luckless officer had gone so far behind enemy lines. A lot of the young pilots on the squadron, according to him, were too keen to go chasing over the lines, thinking they would be "covered in honour and glory". All they achieved, he insisted, was either to "get carbonisé", the French term for being burned to a cinder, or join the Kaiser for a meal in Potsdam.

It seemed that these two flight commanders, at least, were what *La Guerre Aérienne* called "honorary pilots", and hardly an inspiration to their juniors. It was not that there was a cunningly orchestrated attempt to destroy confidence about their conduct. Rather, the impression is of a malevolence amounting to a form of spiritual pollution abroad. Grinnell-Milne was leading a dog's life and one's interpretation is that those who occasioned it were gloating. There was nothing about life as an officer in 16 Squadron to bear out the general reputation of RFC messes as havens of tolerance and concord.

One day Grinnell-Milne attacked an observation balloon a mile behind the German lines. Despite anti-aircraft fire he made two passes at it, his observer punctured it with the Lewis gun, and the enemy had to haul it down. His flight commander accused him of having gone a long way over the lines.

Grinnell-Milne demurred that the balloon was quite near the Front.

This was nonsense, said Wigram. He had seen that balloon himself and it was a long way back. He reminded the offender that he had orders not to cross the lines, and warned the "young feller-me-lad" that he was looking for trouble.

Grinell-Milne returned one day with a damaged propeller. His flight commander demanded to know how the devil he had managed to get two bullet holes in it.

"I had a scrap."

Had the bullets come from the enemy?

Grinnell-Milne replied that he thought so.

Wigram suggested that the shots had come from his own gun. How would he like it if he were made to pay £40 for a new propeller?

Another flight commander, known as Foxy, incurred Dowding's displeasure for being seen, by the wing commander, going round his flight wearing pyjamas under his greatcoat. Foxy said there was nothing wrong with that. Britain would not lose the war because of it. It was his morning in bed instead of dawn patrol. Dowding immediately asked Wing HQ to post this officer away. "No one but this old starched shirt would have made such a fuss about it," Foxy complained. But, he added, of course the Major had to make a fuss, because he was only concerned about missing promotion. Foxy was a regular officer, due for his majority and a squadron of his own, which makes his virulence all the more bitter and exceptional. He must have been very sorely provoked to violate the code by criticising their Commanding Officer to his juniors; mostly holders of mere temporary commissions, at that.

Harmony in a squadron depended on its Commanding Officer. Dowding's was conspicuous for its discord. His unpopularity and mistakes arose from unconquerable shyness as well as from a lugubrious natural austerity. He and Trenchard together could create an atmosphere of excruciating tension. When Trenchard lunched in 16's mess, "Boom", according to Grinnell-Milne, hardly spoke. "Stuffy" droned throughout in a barely audible voice. Everyone else was silent. The pilots had been expecting Trenchard to have a talk with each of them. Instead, he left immediately after lunch. A word would have been encouraging and lack of it made his visit uninspiring.

The contrast between this stultifying atmosphere and that of a good squadron with high morale is exemplifed by comparison with life on 24 Squadron under Lanoe Hawker's command. In 24's mess there was frequent rowdy conviviality. Every week the officers of some other squadron would be invited to dinner and the squadron's hospitality was almost as frequently reciprocated. Most of 24's pilots were twenty years old, some were younger, only Hawker and one other were twenty-five. When at home to guests they fought battles in which armchairs were used as armoured cars, tennis balls as bombs, soda syphons as flame-throwers and fly-sprays as weapons of chemical warfare. These last were also wielded like clubs. The squadron mess was a transportable hut, one wall of which was hinged so that it could be folded for each removal. When a scrum of any sort was in progress, the hosts made a point of manoeuvring it towards this wall. The panel would drop and out would tumble several struggling visitors.

Always concerned about his pilots' recreation, Hawker borrowed horses from the cavalry and had a tennis court marked out. He joined in all activities himself, riding, playing tennis and tumbling about in the roughest of mess games. On one occasion, in another squadron's mess, he was knocked on the head and spent some minutes unconscious. Participation did not lead to familiarity and contempt. He was always respected as well as admired.

Riotous venting of high spirits in the mess in which its CO joined as energetically as his juniors did not impair discipline or efficiency. Within less than three months at the Front, three pilots had been awarded the Military Cross.

Fifty years later, one of his pilots described Hawker as a leader of men who combined modesty with great courage, gentleness with a steely determination, and unselfishness with a most human understanding.

For the Italian airmen, living conditions were pleasanter than on the Western Front. Operations were on a more modest scale because both the opposing air forces were smaller than those engaged in France. Some squadrons were based at established peacetime stations with permanent buildings and on the outskirts of towns where ample recreation was available. Above all, the Italians were at home, unlike the British and Germans; and though the French were also in their own country, their aerodromes were as makeshift as those from which the British and Germans operated.

The Italian Air Corps pilots in particular, and fighter pilots most conspicuously of all, displayed the same jauntiness as in the other warring nations. It was the bomber and reconnaissance squadrons that predominated, so the comparative scarcity of fighter pilots contributed to their standing.

In the first many months after Italy's entry into the war, the Austrian Air Service was the more active and better trained. In addition to constant reconnaissance over the battlefield, it sought to spread panic in the cities and took Italian aerodromes as its objectives. Despite this, Italian airships made many night raids on a variety of targets. The Air Corps was able to give little help to the ground forces fighting the Battles of the Isonzo river, but bombed railway junctions and factories.

The dawn of the year 1916 was severe and harsh on the ground, at sea and in the air. Although the Army found ways to overcome the logistical difficulties of the first winter in the trenches, the Air Corps had to stay grounded by bad weather and low temperatures which damaged the aeroplanes and increased the number of engine failures. But gradually the old machines were replaced by new ones of greater power, and supplies of weapons, bombs, spares and wireless sets began to flow. Up to May 1916, 279 aeroplanes were built.

By the end of March the front-line Air strength was: seven Caproni bomber squadrons; two Voisin and eight Farman reconnaissance squadrons; five Caudron and two Farman artillery observation squadrons; five fighter squadrons; one seaplane squadron.

Provision for aircraft flying over the mountains had to be made. Airstrips were laid out in the valleys for those who had to make forced landings. Telephone and telegraphic communications were established between these and Headquarters.

With an agreeable touch of Latin floridity, the archives tell us that "The fighters were born in the spring of 1916. They were given their baptism of glory by Captain Francesco Baracca, who initiated his series of victories by bringing down two aeroplanes adorned with the enemy's insignia of the black cross, in the sky above Medeuzza.

"For the Mauser pistol, the '91 rifle and the musket, aboard our aeroplanes, were substituted machineguns. At the beginning these were not adapted to use in the air because, apart from being subject to frequent jamming, the magazines held few rounds." Exactly as with the Lewis guns on British and French fighters. It was not until early 1917 that synchronising gear was fitted to Italian fighters. Until then, as in the RFC, machineguns had to be mounted in the nose of a pusher or on the upper mainplane of a tractor.

We should not dismiss Baracca's first success so baldly. It was a great occasion for the entire Italian nation and deserves to be told as his official biographer recounted it. The 7th April 1916 began at 4 o'clock in the morning for No. 1 Fighter Squadron. The rumble of Austrian aeroplanes in the sky resounded in every direction, towards Palmanova, Tricesimo, Casarsa. The red glow of anti-aircraft fire and the shifting bright beams of searchlights had bathed and streaked the sky all night. At first light, one after another aircraft took off, climbing to 2000 metres and dispersing in all directions in search of prey.

After about half an hour had elapsed, Baracca picked out two enemy machines making rapidly towards Gorizia. "I saw above me," the ace recounted in a letter to his family, "the big wings of an Aviatik, with the black cross. He was going fast and I gradually gained on him. When he climbed, I increased speed. Drawing close, I had begun a most difficult manoeuvre to protect me from his shots. I saw the machinegunner aiming in one direction and I veered in another, then vice versa. This game continued for several minutes until I was positioned 50 metres behind his tail at a height of about 3000 metres. Then, in an instant, I aimed and fired 45 rounds. A moment later the enemy swerved heavily to one side and was thrown almost vertical."

The pilot of the Aviatik, a Viennese cadet aged twenty-four, although wounded in the head, with the petrol tank holed, the wings riddled, and the observer slumped over his machinegun, screaming with pain and shedding streams of blood, was able to bring the aeroplane under control and landed in a meadow near Prato. Baracca landed in the same place. A crowd of soldiers threw themselves on him and hoisted him shoulder-high in triumph.

A second enemy fell to Baracca's guns on 16th of May. He scored his third kill in the summer, on 23rd August, and the fourth on 16th September, to end the year with a total of four which rose to thirty-four in the course of three years' fighting.

That engaging character Fulco Ruffo di Calabria achieved his wish to transfer to fighters, in May 1916, when he was sent to a flying school at Cascina Costa on the Gallarate moors, where pilots were instructed on the Nieuport Bébé, which Macchi were manufacturing under licence.

Until that month, the squadrons had been confusingly identified. For each operational function, they were numbered in the same series. Thus there were Nos 1 to 5 Fighter Squadrons, Nos 1 and 2 flying Nieuports and 3–5, Aviatiks. Nos 1–7 Bomber, on Capronis. Nos 5 and 7 Voisin Reconnaissance, 1, 2, 4, 6, and 10–13 Farman Reconnaissance; 1–5 Caudron and 6 and 7 Farman Artillery Observation, and No. 1 Seaplane Squadron. A new system was introduced in May 1916, and the fighter squadron that Ruffo di Calabria joined on 28th July 1916 was No. 77 Fighter. The Air Corps did not have its own uniform. He wore that of his original arm: to which, says his official biographer, in a certain sense he had never ceased to belong.

"Serving in the Air Corps, Ruffo, faithful to the old traditions of his arm, fought with the spirit of the cavalry 'which threw itself against any obstacle', but this quality alone would not have enabled him to add his name to the official roll of aces if it had not been accompanied by exceptional dexterity and aptitude for flying, infallible marksmanship and judgment in attack.

"Also on the physical level he embodied the typical look of a fighter pilot: tall, slender, black-haired and black-eyed, a vivid light in the depths of the pupils, with the sharp profile of a high-flying bird."

On 23rd August he shot down his first enemy aircraft and on 16th September a second. Of the latter engagement there is a detailed description in *Nel Cielo – In The Sky* – one of the few aviation magazines of the time. Part of the article reads: "On 16th September an Austrian artillery observation Löhner was flying between Monte Stol and Monte Starieski, heedless of our anti-aircraft batteries. At the same time two of our fighters were flying in the same area. One of these was Baracca, the other was Ruffo di Calabria, newly converted to the Nieuport. Our airmen dived into the attack. Bursts of machinegun fire were exchanged. The Austrian aeroplane defended itself with consummate mastery. Then a third Nieuport joined in, piloted by Olivari, but by this time all was over for the enemy because Ruffo's machinegun had sent it into a steep bank with a hail of bullets which killed the pilot.

"Because the Löhner, left to its own devices, fell like a dead weight, the Austrian observer tried to take hold of the control column and bring the aircraft under control. But his comrade's body was in the way. It was obvious that the observer was making a desperate attempt to shove the corpse aside and use the joystick, twisting about, pushing and tugging. All was useless. The aeroplane was zig-zagging, out of control. With a supreme effort, the observer got one hand on the control stick. For a moment the machine's fall was arrested. They were at 800 metres and the three Italian fighters were ceaselessly circling, ready to resume firing. But this attempt also failed and the aircraft began to spin, once, twice, three times. Now and again the aeroplane came right for an instant, then began to fall again. At last it hit the ground violently halfway down Monte Stol. The observer was dragged from the charred wreckage. He was mortally wounded, yet managed to find the strength, in hospital, to give an account of his tragic adventure.

"Hardly had Baracca, Ruffo and Olivari landed than they went hastily to the scene of the crash. Meanwhile medical orderlies had laid the pilot's body on a stretcher and others, with a doctor, were giving the wounded man first aid.

" 'Is there any hope?' Baracca asked.

" 'Yes, if we can get him to hospital quickly, perhaps we can save him.'

" 'Poor young fellow,' said Baracca, turning to Ruffo and Olivari, 'but that's war!' "

Much worse was in store. So far, the greatest hazards that airmen had faced on the Italo-Austrian Front were the mountains, which forced them to fly at heights where the cold struck through however many layers of warm clothing they could wear; the peaks that created unpredictable air turbulence; and the lethal terrain awaiting a forced landing. Soon the growing number of aeroplanes on both sides would lead to casualties far higher than the forces of Nature could inflict.

On the Western Front, the imminent butchery on the Somme was about to bring heavy losses also in the air.

CHAPTER 11

———————◆———————

1916. The Crucible

The Somme Offensive was the crucible in whose heat the RFC and the Luftstreitkräfte found their definitive shapes. The series of actions launched by the Allies in an attempt to break through the German lines called for maximum effort by the RFC and Aviation Militaire; which in turn put pressure on the Luftstreitkräfte to exert itself to the utmost. From the outset air activity was greater than at Verdun. The British put 185 aeroplanes into the battle; the French, 200. The long preparatory bombardment, which lasted a whole week, gave the enemy ample warning; they mustered 130 aeroplanes to try to control the air space.

The Allies, then, began with an advantage; which, even before the artillery barrage opened on 1st July, was made all the greater by the removal of their two most formidable opponents: one permanently, the other temporarily. In April, Immelmann, known by now as "The Eagle of Lille", with thirteen kills to his credit, had been granted a regular commission. His brother officers celebrated the event by hiring a band to play at dinner in their mess. Hundreds of troops gathered on the road to listen and to cheer the hero. The Crown Prince of Saxony had decorated him with the Commander's Cross of the Order of St Heinrich, which, being a Saxon, Immelmann rated higher than the Pour le Mérite, a Prussian order.

He was still as unsophisticated as when he joined the Service. He wrote to his mother: "Now I am a full lieutenant and all of a sudden one of the senior comrades. It has been a quick business. I think my career is unparalleled. Only a year ago I was an acting officer without any distinction ... and today!!"

On 18th June, he took off to attack an FE2b of 25 Squadron which had crossed the German lines, flown by Second Lieutenant McCubbin with Corporal Waller as observer. After his first diving pass, Immelmann zoomed into the half-loop that was the first phase of the turn he had invented. As he was about to roll upright at the top, Corporal Waller fired at him. The Fokker broke in two and fell to the ground.

The RFC awarded Waller the kill. The Germans insisted that Immelmann had shot off his own propeller. They said his gun was not synchronised. It had been fitted just before he took off and a new propeller had been bolted on; in the wrong position, they claimed: there had not been time to ensure that the two were in harmony. The matter has never been resolved. A 25 Squadron pilot took the risk of flying over the enemy airfield at fifty feet to drop a wreath "To a gallant and chivalrous opponent".

It was a great compliment to the Vickers FB5, the Gunbus, that in the enemy records the encounter was entered as having been with "a Vickers". So impressed were the Germans with the Gunbus that they used to refer to the DH2 as "a Vickers" and to the FE2 as "a Vickers two-seater biplane".

Immelmann's death was a demoralising shock to both his Service and his nation. Determined that Boelcke, who was now a captain and still in the thick of the fighting, should not be the reason for further damage to pride and confidence, the Kaiser ordered him to be grounded and sent on a public relations tour of the Eastern Front.

A radical reorganisation of the German Air Service, to combat the Allies' air superiority over the Somme, brought him back within two months. The flying units were to be renamed "Jagdstaffeln", literally hunter squadrons – fighter squadrons, in fact – and their establishment was increased to fourteen aircraft. Jagdstaffel No 1 existed only on paper, so the first to be formed was Jagdstaffel (shortened to "Jasta") 2 and Boelcke was given command of it. While visiting the Eastern Front he had renewed acquaintance with Manfred von Richthofen, whom he had met in 1915 in the dining car of a train on which both were travelling on leave.

Boelcke at that time had four victories. Richthofen, who was still an observer, asked him how he did it. His reply was the same as Fonck or Navarre, Ball or Mannock, Rickenbacker or Baracca might have given: "Well, it's quite simple. I fly close to my man and aim well, and then of course he falls down."

"I have done that, Herr Oberleutnant, but my opponents don't go down."

"The reason is that you are in a large machine and I fly a Fokker monoplane."'

Richthofen often recalled those words. On their second meeting he made such a good impression that Boelcke invited him to join Jasta 2.

The FE2b and d, the Gunbus and the Martinsyde Scout, before its relegation to bombing and reconnaissance in October 1915, had shot down many Fokkers, even though they in turn had suffered worse. The Sopwith two-seater 1½-strutter, which the RAF and the RNAS began to receive in 1916, was the first Allied aircraft with a machinegun firing through the propeller arc. Initially, the Vickers-Challenger interrupter gear was fitted for the pilot's gun, but was soon replaced by the Scarff-Dibovsky gear. Originally, also, the Lewis gun for the observer in his rear cockpit was on a Nieuport mounting: which was exchanged for the more efficient Scarff ring. But it was the DH2 that mastered the Fokker on the British Front, while the Nieuport 11 was doing the same in the French sector.

Not until a few months before the Somme Offensive did a British pilot first receive the same adulation in the press as Boelcke, Immelmann, Guynemer, Nungesser and Navarre. Albert Ball was nineteen years old when he joined No. 13 Squadron on 18th February 1916 to fly the BE2c, which was by now outclassed by practically every other aeroplane in the Flanders sky. Here was a young man who epitomised the most dangerous military material: sent out into the world straight from the strict discipline of boarding school, where he had been imbued with obedience, respect for authority, religious faith and the high ideals of bravery, patriotism and honour. It was not the lad who had been hardened by a life of deprivation in the back streets of an industrial city or London's East End, in which he had had to survive by his wits, his resilience, his fists and his boots, who was potentially the most lethal in battle. It was the public school product who was potentially the readiest and most determined killer: instantly acquiescent to the orders of his superiors, made strong and healthy by compulsory games, cross-country runs, cold baths and a sensible diet; with a strong sense of responsibility and an awareness of a privileged upbringing that imposed obligations of leadership and self-sacrifice on him.

At this point it comes instantly to mind that the twenty most successful fighter pilots of the war include several Frenchmen and Germans who had never even heard of the British public school system. Britons such as Mannock and McCudden, and Canadians like Collishaw, Bishop and Barker had never set eyes on a boarding school of any kind. And if the fighting on the Austrian Front had been on a bigger scale, and if

America had entered the war sooner, there would have been Italians and Americans with more than forty victories who were the product of very different systems of education from that of the British middle and upper classes, with its rigorous insistence on unquestioning obedience, Spartan conditions and frequent corporal punishment. Among the average sort of soldier, sailor or airmen, however, the ones who took most easily to obeying orders and putting on a brave face when their bowels were deliquescing with fear, which is the essential requirement in action, were those with a background like Ball's. *Autres temps, autres moeurs*: we are considering a breed of an era long past.

Ball went from Trent College, in Nottinghamshire, into the Sherwood Foresters and transferred to the RFC on 29th January 1916. He had already learned to fly. While in camp on the outskirts of London in 1915, he had taken lessons at Hendon, airborne at first light so as to be back on parade at 6 a.m. Although 13 Squadron's main task was artillery spotting, in April he enabled his observer to shoot down one hostile machine and force down two more. On 7th May he was posted to No. 11 Squadron, which had eight FE2bs, four Vickers Fighters, three Bristol Scouts; and was being re-equipped with Nieuport 11s, of which three had arrived. It was on this last type that he began to score conspicuous successes.

His letters to his parents reveal his immaturity as well as his self-discipline and sense of duty. There is nothing in them to suggest the fighter pilot prowling about the sky like a predatory beast, powerful, menacing, confident. There is nothing in their style, either, that indicates an expensive education. Before leaving England: "I do hope that as you say, I shall come home fit and well, and work hard at my work, but one job at a time is enough for a boy of my age. I simply long to have a smack, but my turn is really a long time coming."

On reaching his first squadron: "At last we are in for the sport and really look like having plenty of it. The machine I am flying is a BE2c, so I do not consider my luck very good, however I shall have a good smack. Oh! I can see heaps of sport ahead, but it really is mad sport."

Three months later: "I must say that although my nerves are quite good, I really do want a rest from all this work. I can stand a lot, but really I have been coming on in leaps and bounds in the last few days, and it is just beginning to tell on me. I always feel tired. I have struck a topping lot of chaps in this squadron and they look after me fine. But they all think me young and call me John. Well, this is no hardship and I am really very happy."

A few days after that, having scored his first solo victory: "Well, I have just come off my patrol on the new machine. You will be pleased to hear that I brought down a Hun Albatross [sic]. He was at 5000 over his lines. I was at 12,000. I dived down at him and put 120 shots into the machine after which he turned over and was completely done in."

In July: "No. 13 has lost four machines and passengers in the past week and our squadron two, and four crashes. However, mad Lonely One is still going strong. They call me John the Lonely One now."

He was already going off on his own to hunt the enemy. His maiden kill was made with a Nieuport 11 and is indicative of his skill. The Albatros was a fearsome recent arrival at the Front. A handful had appeared late the previous year and were now proliferating as a deadly successor to the Fokker. It carried two machineguns firing through the propeller, was faster and a better climber than the Fokker, but less nimble. The Nieuport 11's rate of climb and top speed were superior to the Fokker's. But the Albatros D1 was the most beautiful aeroplane of its generation, as its successive marks were of theirs. The most hackneyed description of it was "sharklike". In fact, it was torpedo shaped, the space between its upper and lower mainplanes was very narrow and it had the sweet, rakish, murderous lines that were un-matched until the Spitfire was seen in the sky more than twenty years later. It is an old axiom in the aviation industry that "if it looks right, it'll fly right". The Albatros amply confirmed this.

A faster Nieuport, the Mk 17, had come into service with the French and a very few had started to reach some British squadrons. No 11 had just received one and Ball was itching to take it up and make his next kill.

Air fighting falls into few categories. Descriptions of air combat become repetitious and tedious to read. Most actions consisted of one killing dive out of sun; a dive that missed and necessitated an immediate zoom with gun firing; a twisting, switchbacking duel between two adversaries; one aeroplane, or a section, outnumbered and fighting off concerted attacks. Ball's favourite method was to stalk his victim patiently, slide up astern and beneath, and despatch him with a short burst from a Lewis gun mounted on the upper wing and pointing obliquely upwards. The machinegun on a Nieuport could be moved by hand on a Foster mounting, named after a sergeant on 11 Squadron who had designed it. This was a quadrant down which the gun was slid to facilitate changing the magazine. It could be returned to the horizontal, pointing dead ahead, or at any upward angle. Ball's canny furtive approach was interestingly in contrast with his frankness of manner and straightforward attitude. It worked extremely well, although its practice evidently imposed considerable nervous tension on him.

His home letters divulge as much about his daily life, intimate emotions and combats as any second party could try to convey by going into details. To his father, 10th July: "You ask me to let the devils have it when I fight. Yes, I always do let them have all I can, but really I don't think them devils. I only scrap because it is my duty, but I do not think anything bad about the Hun. He is just a good chap with very little guts." This is not a judgment that all Allied aircrew would have endorsed. "Nothing makes me feel more rotten than to see them go down, but you can see it is either them or me, so I must do my best to make it a case of *them*."

18th July: "At night I was feeling quite rotten and my nerves were quite poo-poo. Naturally I cannot keep on for ever. So at night I went to see the CO and ask him if I could have a short rest." The Major referred this to the Corps Commander, who ordered Ball detached to No 8 Squadron to fly BE2cs: not only retrogressive after Nieuport 17s, but also highly dangerous. But Ball was lucky. Vindictive, ignorant and cruel though the General was, his failure to understand the mental strain imposed on fighting airmen might have been worse. The usual response to such a request was to send the pilot into the trenches. "This is the thanks after all my work ..." Ball's score had been mounting. "It is a cad's trick."

He soon asked to return to 11. He was posted to No. 60, under Smith Barry, where he cultivated a garden outside his tent. "The peas are topping." He had several crashes. He was awarded the Distinguished Service Order. He wrote on 31st July: "Re rests, I am afraid that they are out of the question, for they don't give you a rest unless you are quite a crock." He forgot that 14th August was to be his twentieth birthday, until his parents sent him a present. Thanking them, he said he would be glad to be home for good.

29th August, to his mother: "A French major called to congratulate me yesterday. He says I have now got more Huns than any pilot out in France. They make it out to be sixteen crashed in eighty-four flights and also a balloon. The major had a long talk with me today. He is very pleased and says I may have leave. Oh! won't it be A1. I do so want to leave all this beastly killing for a time."

On 15th September he was in action again. At 3 p.m. he took off in a Nieuport armed with one fixed Lewis gun on an offensive patrol at 7000 feet. An Albatros, Type A, with guns front and rear, flying at an estimated 80 m.p.h., came in sight. Here is his Combat Report. "Albatros seen going south over Bapaume. Nieuport dived and fired one drum when within 50 yards after which the gun on the Nieuport came down and hit me on the head, preventing me from following the H.A. (hostile aircraft) down." This was a hazard of the Foster mounting when firing at high elevation.

Later that evening he was up again, this time armed with Le Prieur rockets in addition to his Lewis gun. "Five Rolands seen over Bapaume in formation. Nieuport dived and fired rockets in order to break up formation. Formation was lost at once. Nieuport chased nearest machine and got under it, firing one drum at 20 yards. H.A. went down quite out of control and crashed N.E. of Bertincourt."

These reports are made out in the name "Lieut. A. Ball, MC."

Before his next fight, 21st September, his Distinguished Service Order was gazetted, to add to his Military Cross, for the Combat Report of that date is accredited to Lieut. A. Ball, DSO, MC. Decorations came more quickly than in 1939–45.

He met six Rolands flying at about 90 m.p.h. "H.A. seen N. of Bapaume in formation. Nieuport dived and fired rockets. Formation was lost. Nieuport got underneath nearest machine and fired a drum. H.A. dived and landed near railway. Nieuport then attacked another machine and fired two drums from underneath. H.A. went down and was seen to crash at side of railway. After this the rest of the H.A. followed the Nieuport towards the lines and the Nieuport turned and fired remainder of ammunition after which it returned to the aerodrome for more. Second machine was seen to crash by Lieut. Walters."

On 25th September he ran into two formations of Rolands and Type A Albatroses, and saw them both off. "Nieuport could not see any H.A. over Bapaume at a reasonable height, so it went along the Cambrai road. After being there for a few minutes, two formations came along. Nieuport attacked the first. The H.A. ran with noses down, but, when another formation came near it turned towards the Nieuport. The Nieuport fired one drum to scatter the formation after which it turned to change drums. One of the drums dropped into the rudder control and for a few seconds the Nieuport was out of control.

"Nieuport succeeded in getting drum on gun and attacked an Albatros which was then flying at its side. Nieuport fired 90 rounds 1 in 3 Buckingham at about 15 yards range underneath H.A. H.A. went down quite out of control and crashed. The remainder of H.A. followed Nieuport, but in the end left. In order to keep them off at a safe range Nieuport kept turning towards them. Each time this was done H.A. made off with noses down."

Combat at so close a range risked collision, or his own aeroplane catching fire when he set alight to an enemy with Buckingham incendiary rounds.

He had been promoted. This report is by "Capt. A. Ball, DSO, MC."

On 18th September, between whiles, he had written most touchingly to his father: "Oh, you did make my leave a topper, and if I live to be a hundred I shall never wish for a more happy time."

He would not live to be twenty-one.

The land battles on the Somme between July and November 1916 had a direct effect on the concurrent air operations. Air supremacy suddenly shifted once more into German hands. In September the Germans destroyed 123 French and British aeroplanes and lost 27. Boelcke's Jasta 2 was the most effective instrument in changing the balance. Between 17th September, when it first went into action – by which time its commander had amassed 25 victories – and 31st October, it lost a mere 7 machines while bringing down 76 British. Two other Albatros Jastas, Nos 1 and 3, had been formed and were soon operational. In October the total Allied – mostly British – losses were 88; the enemy's, 12. By the next month the number of Jastas had increased to seven.

Jasta 2's operational record got off to a flashing start on its very first patrol. The day was 17th September. The objects of the Jasta's attention were eight BE2cs of 12 Squadron and six escorting Fees of No. 11, on their way to bomb the railway station at Marcoing, deep behind the enemy line. As the antiquated BE2cs waddled in for the attack, the Germans pounced and shot down two of them and two scarcely less vulnerable fighters. Richthofen bagged an FE2b. He had as much cause to take pride in this as a man with a pistol would have for vanquishing a boy with a peashooter. Delighted at what he had accomplished, he landed by the wreck and helped to extricate the mortally injured pilot, Lieutenant L. B. F. Morris, and observer, Lieutenant T. Rees: both of whom died within minutes. In celebration, he wrote to a Berlin jeweller to order a silver cup engraved with the details of his "victory": date, time, place, type of enemy aircraft, name of pilot. He perpetrated the same diseased act of bad taste, in concession to his psychopathic delight in killing, after each of his successes. It was also his morbid habit to scavenge the wreckage of aircraft he had destroyed for souvenirs with which to adorn his mess and a room in his parental home entirely dedicated to this unwholesome display. Some trophies were even, in execrable style, exhibited over his parents' front door. From the site of his first kill in the air, he took the FE2b's machinegun. The whole nasty business reeked of the custom among other savages of decapitating their enemies and shrinking their heads.

The Germans' rejoicing was not unblemished. On 28th October two of Lanoe Hawker's 24 Squadron pilots, Lieutenants Knight and McKay, were far behind enemy lines when Boelcke and his Jasta, Albatros D2s, intercepted them. The Combat Report specifies that twelve Halberstadts and two small Aviatik Scouts attacked the pair of DH2s. The two Britons at once began to circle tightly, which confused the enemy. In the latter's general disorder, the Jasta's oldest member, thirty-seven-year-old Erwin Böhme, whom Boelcke had specially selected, got in Boelcke's way as Boelcke attacked Knight. "After five minutes' strenuous fighting" these two Albatroses collided. Böhme's left wing crashed into Boelcke's right wing and sent Boelcke's aircraft down out of control in a steepening glide which ended in a crash that killed him.

The RFC suffered a comparable blow less than a month later. On 23rd November, Hawker, patrolling at 6000 feet behind the German trenches, with Captain Andrews and Lieutenant Saundby, saw two enemy two-seaters; at which Andrews dived. Let the Combat Report take up the story: " ... and then, seeing two strong hostile patrols approaching high up, was about to retire when Major Hawker dived past him and continued the pursuit.

"The D.H.s were at once attacked by the H.A., one of which dived on to Major Hawker's tail. Captain Andrews drove this machine off, firing 25 rounds at close quarters, but was himself attacked from the rear, and his engine shot through almost immediately, so that he was obliged to try and regain the lines. He last saw Major Hawker engaging one H.A. at about 3000 feet. Lieutenant Saundby having driven one machine off Captain Andrews's tail, engaged a second firing three-quarters of a double drum at 20 yards range.

"The H.A. fell out of control for 1000 feet and then continued to go down vertically. Lieutenant Saundby could then see no other D.H.s, and the H.A. appeared to have moved away east, where they remained for the rest of the patrol."

Richthofen's Combat Report reads: "... with a Vickers single-seater ..." Comment has already been made about the Germans' incorrect aircraft identification. "... The crashed aeroplane lies south of Ligny Sector J. The pilot is dead. Name of pilot: Major Hawker.

"I attacked in company with two aeroplanes of the squadron a single-seater Vickers biplane at about 3000 metres. After a very long circling fight (35 minutes) I had forced down my opponent to 500 metres near Bapaume. He then tried to reach the front, I followed him to 100 metres over Ligny, he fell from this height after 900 shots." The disparity in the heights given by the opponents is noteworthy.

In fact what happened was that Hawker, the far better pilot but in a greatly inferior machine, outflew Richthofen despite the fact that his engine was running roughly from impeded petrol flow which robbed it of full power. Richthofen had to use a huge quantity of ammunition before he finally hit Hawker in the head; his eleventh victim.

Leutnant Stephan Kirmaier succeeded Boelcke in command of the Jasta. Under his leadership it destroyed twenty-five Allied aircraft in twenty-five days. Kirmaier was a sound commander who might have achieved fame if Captain Andrews of 24 Squadron had not shot him down before he was into his stride.

Another member of Jasta 2 laid the foundation of his fame over the Somme. Werner Voss was Jewish, a tailor's son, who had falsified his age to enlist in the hussars. He transferred to the Air Service as an observer and operated as such at the Somme until he became a pilot and joined Boelcke in September. When he made the change he was the only surviving aircrew of those with whom he had begun his operational career. This gave him so sincere a sympathy for the crews of two-seaters that he aimed always for the engine, to give their occupants a chance of survival. Everything that has been said about Voss evokes admiration.

Guynemer was the outstanding French success during this period. Wounds inflicted at Verdun had kept him out of action from March, when he had a total of eight victories, until June. On 16th July he scored his ninth. By the end of November, when the offensive had petered out, they numbered twenty-three.

The squadrons flying what are now known as interception or air superiority fighters were not the only ones embroiled in or drastically affected by the Somme Offensive.

Contact patrolling, low flying in close co-operation with infantry in attack, was a new facet of air operations. Hitherto this had been regarded as wasteful of aircrew and aircraft, and unproductive of accurate information. It had been found that there was no exceptional hazard in flying low; and prearranged signals enabled properly trained air observers to make precise reports on the infantry's progress. Trials with yellow smoke flares on the ground showed that these could be seen at 6000 feet. The French, during their infantry attacks in late 1915, had used flares, signalling lamps and strips of white cloth laid on the ground. In April 1916 Joffre had issued instructions for the use of such means of air-to-ground signalling, and the British had adopted them.

At the Somme, troops laid strips of white cloth on the ground when they reached certain specified points. They also carried metal mirrors on their back packs, which reflected light and could be seen from the air, so that observers could follow the advance. In addition to lamp signalling, ground HQs used panels consisting of six or eight louvred shutters, painted white on one side, which operated rather like a venetian blind. By exposing the white sides, Morse code could be seen, and read by air observers.

Line patrols were flown by pairs of aircraft, to familiarise the ground forces with friendly shapes and markings, and to strafe the enemy. For the first time, fighters – DH2s – were used to clear the air ahead of advancing troops and to tempt enemy aircraft up to fight, giving rise to the term "offensive sweep".

An important ancillary of the main battle in Flanders was strategic bombing. This was aimed at railway lines and junctions, to disrupt the delivery of ammunition and other supplies, at supply and ammunition depots in such places as Lille, Namur, Mons, and at factories in Germany.

Strategic bombing leads us back to Raymond Collishaw, the Canadian who had had to overcome so many obstacles in order to qualify as a pilot in the Royal Naval Air Service, and who became Britain's third-highest-scoring fighter pilot. We left him at the close of 1915, awaiting shipment to England. He sailed from Halifax, Nova Scotia, on 12th January 1916; made his first solo on 16th June, after eight hours and twenty-three minutes dual; flew seven types of aeroplane during training; joined No. 3 Naval Wing on 1st August, to fly the single-seater version of the Sopwith 1½-strutter; and arrived in France, at Luxeuil, on 21st September 1916. The RNAS had been active in France since the war began, but the details of its operations are outside the scope of this work. Certain of them, however, do impinge on it: as will be seen, the RFC had to turn to its sister Service for help when in dire straits.

The wing comprised two squadrons, Red and Blue. A Flight of Red Squadron had five two-seater 1½-strutter bombers and one single-seater fighter. B Flight had the full establishment of five bombers and two fighters; Collishaw flew one of the latter. A Flight of Blue Squadron had four Sopwith bombers, two fighters. B Flight, six Breguet bombers. The wing's operations were directed by Wing Commander Bell Davies, who had won the V.C at the Dardanelles. The Admiralty seemed to know even less about aviation than the War Office: Bell Davies had a strenuous time in France convincing his masters in London that when an aeroplane's engine failed during an operation, it could not be attributed to pilot error nor could punishment be inflicted.

Surprisingly, there were monthly meetings between the Admiralty Air Department and l'Aviation Militaire. At one of these the French had suggested that an Allied bombing squadron should be formed to raid German munition factories. More unexpectedly, the sailors agreed. What was more, they contributed an entire wing. Since the best situation for the wing was in the French sector, it was put under French operational control.

The Naval airmen shared Luxeuil airfield with the Lafayettes and the Quatrième Groupe de Bombardement, commanded by Commandant Félix Happe. This colourful character, over six feet tall with a bushy black beard, parted in the centre, and beetling eyebrows, had a lively sense of humour and was a staunch friend of No. 3 Wing. Among the fighter escadrilles selected to escort the Franco-British bombers was the Escadrille Lafayette. Naturally they and the nautics got on very well. They played baseball against each other and indulged in "some tremendous parties", Collishaw recalled. He also commented that Whiskey, the escadrille's pet lion mascot, "gave newcomers a bit of a start". He says that the publicity given to the Lafayettes was unfair, when there were many more Americans flying with the RFC and RNAS, scoring more kills, winning more decorations, but receiving no public acknowledgment.

The first operation, on 12th October, was on a large scale for the times and modern in conception. The target was the Mauser factory at Oberndorf, 175 miles away. Three Wing was able to put up five bombers and one fighter of A Flight Red Squadron and five bombers and two fighters of B Flight; four bombers and two fighters of A Flight Blue, and six bombers of B Flight. Happe provided twenty bombers. The Escadrille Lafayette and twelve French-built Sopwith 1½-strutters from other escadrilles would escort the raiders as far as their range allowed. Bombers and fighters of the French 7th Army would make a diversionary raid on Lörrach, well to the south of the target and near the Swiss frontier. This was a thoroughly modern stratagem intended to distract enemy fighters.

The Allied formations would take a direct route to target. After bombing, 3 Wing would make for a point north-west of Oberndorf before turning for home at 10,000–12,000 feet. The French would fly home direct. At 1 p.m. a weather reconnaissance – another modern feature – reported favourably. Fifteen minutes later six Farmans of 4th GB took off, followed by A Flight of Red Squadron at half past one and B Flight five minutes afterwards with B Flight of Blue Squadron. At a quarter past two, Blue's A Flight and the remaining French aircraft would depart.

Cloud base descended. The last four bombers were unable to get into formation and turned back. One crashed, but there was no serious injury to the crew.

It was Collishaw's first operation and he said that anyone who claims not to have been nervous on such an occasion has to be an insensitive idiot or to have a bad memory. One of Red B Flight's bombers could not formate, so turned back. The five remaining pilots, Collishaw among them, were all Canadians. Crossing the lines, they met flak, but no one was hit. At 3000 feet three enemy fighters attacked. Collishaw engaged one, his engine cut out after he had fired, he lost 2000 feet and had to return to base. The four bombers of Blue Squadron's A Flight turned back after failing to make formation. Its B Flight found difficulty in climbing to 10,000 feet and did not cross the lines until half past three. By then most of the rest were arriving at Oberndorf. Heavy flak brought one down and its crew was captured. Fighters attacked the remainder, which beat them off. At 4.10 p.m. they thought they were over the target, and bombed; but the town was Donau-Eschingen. One Bréguet was shot down by a fighter and another crash-landed. Both crews were taken prisoner. The Sopwith fighters of Blue Squadron claimed one German fighter destroyed and one probable.

The French lost six bombers. Several fighters crashed on landing. Most of the bomber losses were among those which the Nieuports escorted: the latters' range was too short. Happe thereafter ceased regular daylight raids and resorted to night operations. Three Wing, with its long-range fighters, carried on with daylights.

The raid had scored several hits but caused no serious damage. Three Wing dropped 3900 pounds of bombs, of which not all were on target. Bombing from 10,000 feet with primitive bomb sights could not expect to be accurate.* Total Allied losses were sixteen aircrew killed or made prisoners of war.

On 1st January 1916 *The London Gazette* announced that Sholto Douglas had been mentioned in despatches; on the 14th, the award of a Military Cross to him and Childs, his observer. He went on a week's home leave and thence to 18 Reserve Training Squadron, where he supervised flying training on advanced types, the BE2c and Avro 504J. In May, aged twenty-two, he was posted to Stirling to form a new squadron, No. 43, then take it to France by the end of the year. In August he was moved to Netheravon, where the squadron was supposed to equip with Sopwith 1½-strutter two-seaters armed with a front-firing Vickers for the pilot and a swivel-mounted Lewis for his observer. The squadron's task was to be long-range strategic reconnaissance, which necessitated an aeroplane that could fight as well as do observation.

Douglas was promoted to major, and, on 16th October, attached to No. 70 Squadron at Fienvilliers, to obtain some experience of the work

* Bomber raids by the RAF and Luftwaffe in 1939 and 1940 did not produce much better results.

his own squadron would be doing. No. 70 was carrying out long-range reconnaissance, photographic reconnaissance and offensive patrols. At that time it had the heaviest casualties of the squadrons in France. The aerodrome, where HQ 9th Wing was also situated, was near the village in which Trenchard had his Advanced HQ. The wing was commanded by Lieutenant Colonel Cyril Newall,* but had until recently been under Dowding. When Douglas returned to 43 Squadron at the end of October he found that owing to the casualty rate in France all his pilots and observers, including the flight commanders, had been posted to the Front. He had to begin training all over again with raw material. By January 1917 the squadron was fully equipped with 1½-strutters, and on 17th January went to France.

If observers are in a neglected aircrew category, there is another which has received even less attention. In those two-seater squadrons that did not need observers trained in reconnaissance or photography, the man who accompanied the pilot had to be a trained machinegunner. He wore an observer's badge and was usually called an observer, but the term "aerial gunner" was also used.

One such was Arch Whitehouse, born in England on 11th December 1885. His parents took him to the USA ten years later but he remained a British subject. At the age of fourteen he left school for a poorly paid job. When Britain declared war he managed, in the face of many difficulties, to return home via Canada and join the Northamptonshire Imperial Yeomanry.

Because he proved a good shot he was sent on a machinegun course. Late in 1915 he landed in France. The regiment handed its chargers over to the 4th Punjab Horse and he found himself helping to look after a hundred mules. He escaped from this by going on another machinegun course, on his return from which he was set to guarding prisoners of war. Shortly after, he was grooming horses and "living like a pig" in mud and manure. An enemy aircraft flew over one day, chased by a British scout which shot it down in flames. Here, young Whitehouse told himself, was a way out of this miserable life: he would fly. The Somme Offensive had begun, volunteer aerial gunners were wanted. Ten days later he passed the medical exam and expected to be sent home for training. Instead, he was told to report to No. 22 Squadron, which flew FE2bs.

He claimed that, rejoicing, he threw his rifle, bayonet, steel helmet (which had just been introduced) and spurs into the River Somme. This

---

* Later Air Chief Marshal Sir Cyril Newall, GCB, OM, GCMG, CBE, AM, Chief of Air Staff.

defies credulity. A soldier's equipment was on his charge and his rifle was the most precious item with which he was entrusted. To lose any article was a serious offence. To lose a rifle meant a court martial and a severe sentence. To throw it away would have earned six months in a military prison, at least. On being posted, Whitehouse must have had to return his rifle and the rest to Stores.

His reception by his new comrades gave him small cause for glee. As he entered the aerodrome a sergeant asked what he was doing there.

Whitehouse replied that he had been up the line as a machinegunner and was now reporting as a volunteer to fly.

The sergeant told him that aerial gunners were barmy and sent him to the Orderly Room. There, the sergeant major's comment was that anybody who would crawl out of the trenches to become an aerial gunner must be a bloody fool. He issued Whitehouse with a flying helmet, leather coat, goggles, flying boots and gloves: the coat stained with the blood of its previous wearer, a commissioned observer who had been killed. Then Whitehouse drew his new uniform and hurried off to don it.

He was told that when he had flown fifty hours over the lines and passed his ground tests he would be given his observer's badge. Then he would be eligible for pilot training in England and a commission. Meanwhile his four shillings a day flying pay would begin at once. As a Second Class Air Mechanic this would make his weekly pay one pound eighteen shillings and sixpence: far more than he had earned as a trooper or in his wretched civilian job. He was delighted.

Hearing an aero engine ticking over near his barrack hut, he went to the door. A pilot climbed down from a Fee. "You Whitehouse, the new gunner?"

"Yes, sir, but ..."

"Good! Get your gear on. I'm going on an engine test."

"I've never been up, sir."

"Fine. Nothing to unlearn."

The pilot, Captain Clement, a Canadian, told him he could sit on the floor until they reached the balloon line. Then he could get up and perch on the spare parts locker or the edge of the nacelle.

Whitehouse confesses that he was cold and scared, and quaked with fear when Clement crossed the enemy lines at 8000 feet. They came under anti-aircraft fire and Clement threw the Fee about in evasion. The stink of smoke from the shell bursts was sickening.

They returned to their own side of the lines and Clement shouted to him to prepare to fire at a ground target, a white wing panel. The aircraft dived and Whitehouse tumbled about in his cockpit. Eventually he managed to fire his front gun and hit the target. Clement's praise seems to have been extravagant: "Holy Smoke! They certainly train gunners in that Yeomanry outfit. Keep this up, young fellow, and you can fly with me any day." These were the first kind words addressed to Whitehouse since he put on khaki, and banished his fear of flying.

He went to his hut to make his bed, but was summoned at once to go on patrol with the rest of C Flight. This time his pilot was a Lieutenant Brooks. They made 8000 feet over the field and formated. Cowering down in his cockpit, he saw that the other gunners were nonchalantly sitting on the rims of theirs. Reluctantly, he followed suit. They were soon under anti-aircraft fire. A few minutes later tracer bullets came darting at them and he saw enemy aircraft. The other gunners and he opened fire. The enemy drew off. The Fees wheeled about.

More hostiles appeared. He straddled the cockpit and stood up to use the gun at his rear on its telescopic mount. He fired several short bursts. A burning aeroplane swept past, enveloping him in smoke and throwing débris against the Fee. He, who had been doubting that he could survive fifty operational sorties, had shot down an enemy machine on his first one. He had joined the squadron less than five hours ago.

The next morning C Flight took off early. He flew again with Clement, who informed him that they were going on an ordinary offensive patrol. They would cross at a point opposite Cambrai and "work the Jerry line" all the way down to St Quentin. On the way back they would do their regular Sunday show: balloons.

They saw Nieuports and Sopwiths returning from dawn patrol, RE8s and BE2cs beginning their artillery-spotting chores. Flak opened up. In the distance the minute shapes of enemy aircraft flitted about, waiting for the RFC's offensive patrols. Clement rocked his wings to attract Whitehouse's attention and pointed down at three German machines. The Fees began to circle above them, firing bursts. One enemy aeroplane rolled on its back, a wing fell off, smoke and flames belched from it. Clement yelled: "Your first burst hacked out a wing root."

C Flight pressed on further behind the German trenches. Shells burst around them, spreading acrid black smoke. Whitehouse heard a tremendous explosion and saw an FE2b fall to pieces. One of the crew tumbled out, arms and legs thrashing and kicking. The wreckage fell onto another FE2b, which began to break up as the tangled mess spun earthwards, burning. Its gunner tried to heave himself onto a wing but was engulfed in flames.

Clement hammered on the fuselage between them and Whitehouse saw another enemy aeroplane ahead. It darted from side to side as he

[151]

shot at it. A wing cracked and folded back, then the other wing snapped off and it followed the two Fees down, in flames.

There were still four of C Flight intact. They held diamond formation and continued the patrol.

Whitehouse allegedly completed his fifty operational flights in three weeks; but with aerial battles over the Somme voraciously consuming aeroplanes and lives there was no question of his being spared to go home and take a pilot's course. He was, however, promoted to AM1, Air Mechanic First Class, and his pay went up to five shillings and sixpence a day.

The Allies gave their ground forces intensive support by contact patrols and low-level strafing. For the pilots it became, in the final months of the war, the most hated and feared form of aerial aggression. Low-level attack instilled a feeling of defencelessness into infantry in their trenches. The words of a German officer convey very well what it was like to have to endure this type of assault. "The infantry had no training in defence against very low-flying aircraft. Moreover, they had no confidence in their ability to shoot these machines down if they were determined to press home their attacks. As a result, they were seized with a fear amounting almost to panic; a fear that was fostered by the incessant activity and hostility of enemy aeroplanes."

The diary of a German prisoner of war confirms this. "During the day one hardly dares to be seen in the trench owing to the English aeroplanes. They fly so low that it is a wonder they do not pull one out of the trench. Nothing is to be seen of our heroic German airmen. One can hardly calculate how much additional loss of life and strain on the nerves this costs us."

And an unfinished letter found on the body of its writer: "We are in reserve but cannot remain long on account of hostile aircraft. About our own aeroplanes one must be almost too ashamed to write. It is simply scandalous. They fly as far as this village but no further, whereas the English are always flying over our lines, directing artillery shoots, thereby getting all their shells right into our trenches. This moral defeat has a bad effect on us all."

The superior performance of the Allied fighters at that time enabled their pilots to use their qualities of courage and skill to the full and win the dominance in the air that is essential for such activities.

The exploits of a pilot on No. 60 Squadron, Second Lieutenant C. A. Ridley, offer a measure of light relief amid the grimness of the Somme. His mission on 3rd August was to drop a French spy behind the German lines. Engine trouble forced them down; prematurely, but on enemy territory. For more than three weeks they managed to hide and to make their way towards Belgium, where they parted. Ridley spoke neither French nor German. Having obtained civilian clothes, he was in danger of being shot as a spy if caught. With brilliant imaginativeness, he bandaged his head, painted his face with iodine and pretended to be a deaf mute. After various misadventures a suspicious military policeman arrested him on a train. Waiting until it had slowed to some 15 m.p.h., he knocked out his captor and jumped out. After further distressing experiences he found a friendly Belgian who helped him to put a ladder against the electrified fence at the Dutch frontier and climb over. It was then 8th October. One week later he reported back to his squadron: bearing a vast amount of invaluable intelligence about German aerodromes, ammunition dumps, troop concentrations and movements.

All any man can do is to try to adjust himself within the limits of constantly changing circumstances. Ridley did better than most.

CHAPTER 12

———————◆———————

1916. Beyond the Somme

Baron Elard von Loewenstern, who had embraced the airman's life with such glee and been despatched to the Eastern Front when he completed his training as an observer, had a lot to say about the Western Front, where he arrived in the summer of 1916. There, he said, at the beginning, before the war became static, the opposing air forces massed twenty kilometres deep behind both sides of the front line. From the forward trenches to twenty kilometres beyond, the sky was full of airmen. Note, not "aeroplanes", but "Flieger . . . flyers". He obviously personalised it to emphasise his pride and pleasure in being one of this company. Here it was normal to greet one another "almost jovially", according to him. "The German flying units had to go about their duties in a sky crammed with enemy aircraft."

He was not there and this is hyperbolical. Neither side had enough aeroplanes to be able to fill the sky with them. Enemies waved to each other in resignation that they could do nothing aggressive, rather than in genial greeting. And, from the outset, the German fighter pilots preferred to stay behind their own lines, while it was the British and French fighters that went far over enemy ground.

This was true throughout the war. It was always the German fighter pilots' policy to stay on their own side of the lines and let the Allies come to them. "Let the customers come to the shop," Richthofen used to say. The Germans knew they could do this because the French and British would be ashamed to hang back. Later, Barès and Trenchard made sure that they did spend all their airborne time on the German side, whatever the cost. Dowding would not have been so callous. He knew what it was like to fly in the battle zone. He didn't prate to the squadrons on his wing that he wouldn't ask them to do anything he wouldn't do: he went up and did it. As a wing commander he used to fly over the Somme. On one occasion a machinegun bullet nearly took his hand off at the wrist when it pierced his joystick; which he kept as a memento. The prevailing wind from the west favoured the Germans' tactic. The Allied aircraft had to fly home in the teeth of it. Their enemies, with the wind behind them, were able to catch them up. Enemy anti-aircraft batteries' aim was made easier because the targets were moving slowly.

It was obvious, said Loewenstern, that the squadrons facing the Germans on the Western Front were much more efficient than the Russians: who confined themselves mostly to defence. So did his own countrymen in the West, but he does not seem to have been aware of it.

The Russian fighter pilots were not highly dangerous, in his opinion. They were not aggressive. The Germans had an easy time in the air. Most reconnaissances could be flown at heights from which the ground was easily scanned with the naked eye. They were within range of the ground defences, but the Russian batteries remained stationary, so one was able to dodge them. Reconnaissances flown at the greater altitudes which were necessary on the Western Front were much more difficult. One had to be alert "not to enter the land of fantasy" and "see" fortifications, troop concentrations, tanks and so forth. (The British had used tanks for the first time towards the end of the Somme Offensive.) A reliable visual search could be made from 2000 metres. This confirms what the RFC had found. "The doubled tension of doing one's job while under anti-aircraft fire quickly wore out one's nerves."

"Hatte man in Osten geglaubt, grosse Leistungen vollbracht zu haben, so musst man – nach dem Westen versetzt – sehr bald erkennen, dass das Fliegen im Osten dagegen heinahe ein Kinderspiel gewesen war." "If anyone believed he had accomplished great performances in the East, he must – after being transferred to the West – very soon realise that in comparison flying in the East had been close to child's play." He continues: "I was terrified, in the truest sense of the word, when on my first introduction in 1916 I saw a fighter appear suddenly out of the clear bright heaven and the glittering tracks of tracer fire all around."

He could have found it an even greater strain on the nerves if he had been flying for the Allies; although the French anti-aircraft gunners, in the tradition of the artillery, which was their Army's pride, were as formidable as the Germans'. From the start of hostilities, the Luft-streitkräfte had dreaded them, and pilots who had faced it warned their comrades of "the hell that awaited them when they crossed the French lines".

The British anti-aircraft defence, however, was poorly equipped and scanty. Initially, the Royal Artillery had no guns specifically designed for the purpose. They had to rely on pompoms firing one-pound shells; which had no demoralising effect, because they did not burst in the air. Thirteen-pounder field guns were adapted. Mounted on motor vehicles, they fired either high explosive or shrapnel, but only to a modest height. By January 1915 each Division was being provided with two of these: a miserly provision. That summer, firing had to be limited to shrapnel because high-explosive shells were sometimes bursting prematurely in the gun barrels. Shrapnel was the more effective against unarmoured aircraft, but only with a direct hit. It did not burst with the loud noise of high explosive, which in itself was frightening and demoralising. Also, HE shells did not have to score a direct hit: their splinters were enough to destroy any aeroplane.

By the end of July 1915 there had been 28 Divisions in France, but only 13 pairs of A-A guns. In 1916, 18-pound field guns were bored out to 3 inches calibre to replace the 13-pounders. They were also mobile but had a ceiling of only 19,000 feet. German guns were effective well above this height. In 1916 Haig told the War Office that of the 112 guns he needed, he had only 67. By the start of the Somme he had 113, of which 70 were the old 13-pounders, plus 12 2-pound pompoms. Loewenstern, when he flew over the British sector, was luckier than he realised, however much of a pounding his nervous system was taking.

Elard von Loewenstern's equable temperament was matched by the placid, sanguine character of Haupt-Heydemarck, whose first airborne experience had been the hair-raising flight in a balloon which ended in a crash that left him with one leg shorter than the other, but his morale and sense of humour unimpaired. In the spring of 1916 he had been posted as an observer to Flieger Abteilung 17, a long-range reconnaissance unit, at Attigny.

He was crewed with a pilot named Engmann, nicknamed Take, and his description of their first operation together could have been recounted by almost any of his British or French counterparts. In its staccato way it contains all the elements of bleary-eyed, reluctant waking, delays, frustration, fear, exhilaration and tenacity which were familiar to all operational airmen.

When his batman woke him on the morning of his first flight against the enemy, his immediate question was about the weather, which was starlit but cloudy. A quarter of an hour later he joined "Take" Engmann at breakfast: "Boiling hot coffee, with bread and marmalade and a lightly boiled egg. This was rare, a privilege for the crew that had the morning's first sortie."

They studied the map together, tracing the track that they must make good in order to photograph railway tracks at Chalons-West, supply dumps south of Buffy-le-Château, and aerodromes at Courtifols, Tilloy-Bellay and Auve. A westerly wind was forecast. Out of doors, they saw that the clouds were drifting north-east, which meant a south-westerly wind. Engmann predicted that there would soon be a lot more cloud "with a lot of muck behind it".

They did their external checks of the aircraft, dismissed the ground crew and returned to snatch some more sleep, fully dressed, until the weather improved. At 6.20 a.m. their batmen woke them with the information that a patch of blue sky had appeared. Ten minutes later, satisfied with a reported north-west wind of 36 kilometres an hour at 3000 metres, they were ready to take off. After another half hour they were at their operational height, with a clear view. The altimeter settled at 2100 metres. Haupt-Heydemarck set the shutter speed at 1/250 of a second and they were ready to begin photographing.

To order a turn he would tap his pilot on the right or left shoulder. A prod between the shoulderblades meant straight ahead. A rap on the crash helmet: below. A wave with the flat of the hand: enemy aircraft. Clenched fist: flak.

When Engmann wanted to warn him he blipped the throttle.

When they had crossed the Allied lines, "The tension to which our nerves were increasingly subjected during the flight reached its apogee: when will the anti-aircraft guns open fire and from where?"

Engmann wriggled on his seat above the petrol tank. He constantly wove to right and left so that he could look ahead past the engine. Haupt-Heydemarck kept stretching his neck to peer over the side and scan the air and landscape.

They saw the trenches slowly disappear beneath them. It seemed endlessly slow, yet they were moving at a speed at 120 k.p.h. Still no shooting. There was no need to give the experienced Take helpful taps: he knew how to make a difficult target for the artillerymen below.

There was a loud explosion and a blow smote the aeroplane as though struck by a wind squall: the first shell bursting, and it was well placed. Engmann instinctively banked away in a steep right turn and the second shell burst far to the left. They did not hold their new direction long. "Half left. Now we'll see where the next shell bursts. Aha, that's good: a couple of hundred metres on our left is a cloud." But now the battery commander had about the right direction. Two shells burst ahead. They sideslipped. It needed only a shell splinter in the belly or shrapnel bullet through the head, "and curtains!" They caught the stench of high explosive. They climbed to 3000 metres. Shells burst on both sides, then one very close on the left and another astern. Engmann made a steep diving turn to the left.

They found their objectives, came under fire over each, but took their photographs. On the way home, passing under the fringes of a cloud, Haupt-Heydemarck looked up: about twenty metres overhead was a Nieuport. It must turn before it could attack. But Haupt-Heydemarck had already trained his machinegun on it and begun firing a few short bursts. The Nieuport wheeled. Haupt-Heydemarck heard 'tack-atackatackatack' on his left and turned the gun in that direction. "Ah, another Frenchman. And – damn it! – a little further behind, a third."

The third Nieuport was passing so that it could try to take the Germans unawares later. But the second one was unpleasantly close. Haupt-Heydemarck would take it first. Carefully he got him in his sights, "Then I let him have it." But the Frenchman was alert and dodged aside, to come curving in again. Number three was coming in from the rear. Number one at the same time from the right.

This last was the nearest and must be dealt with first. Haupt-Heydemarck fired two ten-shot bursts. That seemed to dampen the French pilot's aggression somewhat. He made a wide circle and stayed well away. Once more a fresh burst of fire crackled down from above. With his machinegun that swivelled in any direction, Haupt-Hendemarck could almost return the fire simultaneously. "Enough firing – twelve rounds – eyes up – press the trigger – short pause – change aim and shoot. Always thinking: be economical with the ammunition! Wonderful after all how calm one is. No trace of agitation, no change in my pulse rate." This is unlikely: exertion in the thin air would have made his heart accelerate. Hectic activity coupled with fear must have made it race.

The adversaries were 100 metres apart. Engmann swung towards one of the others. This tactic took Haupt-Heydemarck unawares. As he tried to get another target in his sights he heard a metallic clatter on his right. This time he fired a thirty-second burst, and "hurrah! the fighter slumped onto its left wing, it stood on its head and down it went".

Hardly had the next one, from 200 metres range, seen what had happened, than it veered away to a wary distance. A moment later a shell burst between the Germans and the other two French aeroplanes, which turned and fled southward. It was French AA fire, loosed off with dashing Gallic disregard for the presence of two of their own aircraft. Haupt-Heydemarck and Engmann went home too.

When they landed, the Commanding Officer emerged from the control hut and shook their hands. "Well?"

Haupt-Heydemarck reported that railway traffic was trifling, there was normal activity at the railway stations and aerodromes. Flak en route and over Chalons was heavy, "otherwise nothing special". He gave a short account of their air fight.

Meanwhile Engmann had examined the aircraft and found six holes: two flak hits, four from bullets in air combat.

In July, of twelve crews, the unit lost fifty per cent: two crews killed; one lightly and three heavily wounded.

A fellow victim of severe injuries, this time in an aeroplane accident, was Leo Leonhardn, who was destined to earn the name "der eiserne Kommandeur", "the Iron Commander". In 1915 he was commanding Flight Section 25. By late 1916, with the rank of Hauptmann (Captain), he had been given command of Bombengeschwader 6. His address to his officers and men, when taking over, included three short sentences that left no need to say more: "Duty is everything. Praise I know not. I reprimand only when I see duty set aside." His tough words were accepted and effective. Many, recalling them, attributed the splendid successes of the Geschwader to "the way he led along a rocky, thorny path". His energy was extraordinary, his uncompromising nature was flavoured with a sense of humour. "Only by the strictest observance of his motto, 'Iron, stubborn and impossible to divert', could he and his comrades have achieved the highest possible performance under the most difficult conditions. No operational flight was undertaken without Leonhardn, this half crippled man, conscious of duty, placing himself where he could protect his men from the enemy." That is what the record says, and no one could wish for higher praise.

The battles at Verdun and on the Somme had no direct effect on the Italian Front. The First Battle of the Isonzo, from 23rd June to 7th July 1915, with which the Italo-Austrian campaign opened, however, was the start of a similar series of engagements between entrenched opponents. Both sides were handicapped by the nature of the terrain. Venezia, the Italian frontier province, was a salient with Austrian Trentino on its north and the Adriatic Sea on its south. Between the sea and the mountains lay lower ground where the River Isonzo flowed from the Julian Alps to the Bay of Trieste. It was Italy that attacked and Austria that defended. In the six months during which the first nine Battles of the Isonzo were fought, there were 240,000 Italian casualties and rather more than half as many Austrian.

Co-ordinating the Italian air effort with the needs of the land forces had to proceed slowly for lack of aircraft. Scope lay only in bombing:

strategic bombing, because there were no dramatic surges forward on the ground which could be backed up by tactical bombing or close-support strafing. And there was a lot for the bombers to do. Italy was not engaged only along its frontier with Austria in the Dolomites. There were targets to be bombed in former Italian territory that had fallen into Austrian hands; others in what is now known as Yugoslavia, and in Albania.

It is surprising that so little attention had been paid to fighters. The impetuous, romantic, fiery Latin temperament of the Italian male would seem to be best adapted to the aerial duels that fighter operations used to be from 1914 until the 1950s and the Korean War. It is also surprising that Italian engineers, with their flair for creating beautiful artefacts, were not called on to design small, sleek, swift aircraft. Instead, Italy continued to manufacture French aeroplanes under licence. Favouring bombers, there was the fact that Italy had flown the first military operational sorties in history, which had been to bomb the enemy. So there was a tradition, however brief, of bombing.

The official history of Italian bomber operations in 1915–18 has a chapter headed "Bombing, the Essence of the Air Arm". General Felice Porro, in his account of the Italian Air Corps in the Great War, wrote: "The bomber arm, which had the pride of flying an aeroplane that was clearly Italian in conception and design, which had the honour and glory of carrying the tricolour insignia on its wings in the skies of France, could not however attain that progressive development which would have been desirable and so necessary." This was the Caproni series. All were three-engined, with a seventy-three-foot wingspan, central nacelle and twin-boom fuselage. Of the three marks built between 1913 and 1916, the Ca1 had 100-h.p. Fiat engines; the Ca2, two of these and one 150-h.p. Isotta-Fraschini; the Ca3, three of the latter.

In June 1916 the first large-scale air raid was made when thirty-four Caproni, not in formation but in quick succession, bombed the Austrian aerodrome at Pergine. On 1st August ten bombed Fiume dockyard.

Brigadier General Armando Armani, who, as Captain, Major and Lieutenant Colonel, flew bombers throughout the war "assiduamente e valorosamente", "assiduously and bravely", described an encounter between a Caproni and two Austrian fighters, which "suscitó molta emozione in Italia", "roused much feeling in Italy". (Not a matter of great difficulty.)

"On the morning of 18th February 1916 Caproni 300 (Ca1) of 300 h.p., No. 478, 'Aquila Romana', 'Roman Eagle', piloted by Captain of Artillery Luigi Bailo and Administrative Captain Oreste Salomone, with Lieutenant Colonel of Artillery Alfredo Barbieri, commanding the Air

Lieutenant General Sir David Henderson, first Commander of the Royal Flying Corps and "grandfather" of the Royal Air Force.

Staff of Central Flying School, 1913. Standing: Captain Lithgow, Medical Officer. Assistant Paymaster Lidderdale. Major H. M. Trenchard, DSO, Assistant Commandant (later Marshal of the RAF, Lord Trenchard, "father" of the RAF). Captain Paine, Commandant. Honorary Lieutenant Kirby, VC, Quartermaster. Engineer Lieutenant Randall. Seated: Flying instructors: Captain Fulton, Lieutenant Longmore, Captain Salmond, Major Gerrard.

BE2, the RFC's main aircraft at the outbreak of war. Lieutenant H. D. Harvey-Kelly, the first RFC pilot to land in France, is shown resting during his flight from No. 2 Squadron's base at Montrose to Dover.

Leutnant Max Immelmann and wreckage of British machine he shot down.

Leutnant (later Hauptmann) Oswald Boelcke, Germany's first great fighter leader. Commanded the first Jagdstaffel (fighter squadron).

Vickers FB5 ''Gunbus'', flown by the first homogeneously equipped RFC squadron, No. 11, in 1915; and regarded as the world's first purpose-designed fighter.

VFB5 ''Gunbus'' in flight.

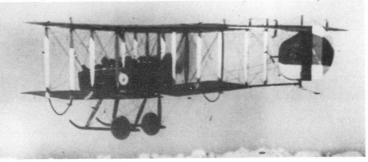

Major Lanoe Hawker, the RFC's third winner of the Victoria Cross and epitome of the British gentleman. A better pilot and shot than Richthofen; who, in a long fight, could not kill him until Hawker's engine faltered.

Gilbert Insall, who won the RFC's fifth VC in an unusual way.

Lieutenant Colonel (later Général de Division) Joseph Edouard Barès, Senior Air Staff Officer at French General Headquarters, September 1914 to February 1917.

Commandant de Rose, France's outstanding commander and tactical organiser of fighter units.

Commandant Antonin Brocard, first Commanding Officer of Les Cigognes, l'Aviation Militaire's mos distinguished fighter unit

Left: Maggiore Pier Ruggiero Piccio (twenty-four victories), Italy's Number Three. Right: Capitano Fulco Ruffo di Calabria, seventh Duke of Guardia Lombarda, who brought down twenty enemy machines and ranked fifth.

Italy's most famous poet, in 1915. Tenente (later Tenente Colonnello) Gabriele d'Annunzio, observer, on right, with his pilot, Capitano Beltramo.

Fixed camera,
operated by
pilot; and
observer with
hand-held
camera.

Making a mosaic map
with photographs taken
behind enemy lines.

Pilots of No. 15
Squadron reporting to
Commanding Officer,
Major H. V. Stammers,
on return from patrol.

RE8 ("Harry Tate"), the most widely used British two-seater on the Western Front: for artillery observation, reconnaissance, ground-support, night bombing.

DH4, the first aircraft designed for high speed day bombing, 1917. Note squadron markings and, on fin, pilot's personal motif.

The British fighter that shot down the greatest number of enemy aircraft. A Sopwith Camel, carrying twenty-pound bombs for ground strafing, in racks between wheel struts.

No. 56 Squadron pilots before
departure to France, 1917.
Standing: Lieutenants G. C.
Maxwell, W. B. Melville,
H. M. T. Lehmann, MC.
C. R. W. Knight, L. M. Barlow,
K. J. Kaggs. Seated: Lieutenants
C. A. Lewis MC (author of
Sagittarius Rising, a classic war
reminiscence) and J. O. Leach
MC, Major R. G. Blomfield,
Captain A. Ball, DSO, MC (later
also VC), Lieutenant R. T. C.
Hoidge.

Captain Albert Ball, VC, DSO,
MC, in his SE5.

SE5a cockpit.

Major Edward "Mick" Mannock, VC, DSO, MC. Top-scoring British fighter pilot: seventy-three kills.

Lieutenant Colonel W. A. Bishop, VC, DSO, MC, a Canadian who came second to Mannock, with only one less victory, seventy-two.

Major J. T. B. McCudden, VC, DSO, MC, MM, who scored fifty-seven victories and ranked fourth among British fighter aces, with his dog, "Bruiser".

Captain A. W. Beauchamp-Proctor, VC, DSO, MC, the South African who scored fifty-four victories.

◼ ANNOUNCEMENTS

BIRTHS

CARINGTON.—On July 24, to DANIELA (née Diotallevi) and RUPERT, a daughter (Francesca). a sister for Robert.

CHANIN.—On July 21, to CLARE (née Thyne) and PAUL, a son (Christopher Thomas William).

GROCOTT.—On July 21, 1993, at Basingstoke, to JULIET (née Haviland) and CHARLES, a daughter (Evangeline Phillippa Hope).

KEILY.—On July 22, in South Africa, to KIM and SEAN, a daughter, sister for Max and Dom.

O'LOUGHLIN.—On June 30, at Heatherwood, Ascot, to SUSAN (née Stephenson) and MICHAEL, a daughter (Amelia 'Amy' Louise).

PRITCHARD.—On July 22, to CAROLYN (née Furby) and MARTIN, a son (Christopher Martin Gordon).

STIRLING.—On July 16, 1993, to PLUM and ARCHIE, a daughter (Amy Sophia Marjory).

WRIGHT.—On July 20, to LISA (née Groves) and JOHNNIE, a daughter (Phoebe Gudrun), a sister for Finnian Henry.

GOLDEN WEDDINGS

BRANGWIN - WEBB.—On July 26, 1943, at H.M.S. Cochrane, Rosyth, BRANGS to PHYLLIS.

DIAMOND WEDDINGS

CLEWS - SIMS.—On July 26, 1933, at the Parish Church of Holy Trinity, Sutton Coldfield, Warwickshire, DENIS to DOROTHY. A family reception celebrates the anniversary.

PHILLIPS - GARDNER-POWELL.—On July 26, 1933, at Birmingham, NATHAN to DOREEN. Now at Canford Cliffs, Dorset. Congratulations from all who love you.

IN MEMORIAM

"THEIR NAME LIVETH FOR EVERMORE"

MANNOCK.—Major EDWARD (Mick) MANNOCK RFC, VC, DSO, MC. 'King of Aces' no. 85 Squadron. Pacaut Wood, France. July 26, 1918. "He faced the void without fear to fly in other skies." A.F.S.

DEATHS

BANDY.—On July 22, 1993, peacefully after a short illness, STANLEY, aged 86 years of Chelford, Cheshire (ret'd Division Manager Pearl Assurance). The loved and loving husband of Kathleen, much loved father of Robin,

MAUDE.—On July 21, 1993, peacefully at home, HESTER JOAN, aged 90 years, widow of Brigadier KIT MAUDE, much loved mother of Gillian, Priscilla and Elizabeth, and grandmother of Alexandra, Rupert, Andrew, Victoria, William and Ben. Funeral private. Family flowers only. Service of Thanksgiving at All Saints' Church, Old Heathfield at 3 p.m. on Thursday, July 29.

MAY.—On July 23, at Horncastle House Nursing Home, DOUGLAS OSBORNE, in his 90th year, formerly of Bolney and in retirement Northern Cyprus, fifty two years with Killby and Gayford. Beloved husband of the late Vi and much loved father, stepfather, grandfather and great grandfather. Funeral at St Mary Magdalene, Bolney on Friday, July 30 at 11 a.m. No flowers. Donations to Royal British Legion, c/o William Collins and Son, 12 Mill Road, Burgess Hill, Sussex, tel. 0444 871515.

MORTIMER.—On July 21, 1993, peacefully at Milverton Nursing Home, Surbiton, JOAN KATHLEEN TERESA (Terry). Formerly of Gravesend and of Royal Dutch Shell. Loving wife of the late Eric. Will be greatly missed by her many friends and relatives. Funeral St Mark's Church, Surbiton, 2.45 p.m, Friday, July 30, followed by cremation at Ballards Park, Leatherhead. Family flowers only, donations to R.N.L.I. Inquiries to Frederick Paine, 081-399 2060.

NEWTON.—On July 18, suddenly, ANDREW DAVID ROBERT, aged 24, beloved son of Douglas and Joan and much loved brother of Karen and Tracey. Funeral service, Thursday, July 29 at 1.30 p.m. at Leigh Parish Church (nr Sherborne). Family flowers only please. Donations, if desired, to Leukaemia Research, c/o W.S. Brister & Son, Hound Street, Sherborne, Dorset (0935) 812647.

REASON.—On July 22, in Goldsborough Nursing Home, Enfield, ERIC FRANK, loving husband of Violet, father of Janet and David. Funeral at Enfield Crematorim on Monday, Aug. 2 at 1.30 p.m. Flower sprays to Co-operative Funeral Services, 113 Lancaster Road, EN2 0JN by 11.30 a.m.

REID.—On July 23, ANTHONY REID, aged 88, beloved husband of Pearl, of Seven Kings, Ilford, Essex. Funeral service at The City of London Crematorium, Manor Park on Thursday, July 29 at 4.30 p.m. No flowers at family request. All inquiries to J Cooper (Undertakers) Limited, 3 Station Parade, Barking, Essex IG11 8ED, tel. 081 594 2339.

COMMENTARY

W.F.DEEDES

Pity about that 'far-flung stuff'

LOCAL NEWS is preferred to national, and national to international. This finding from the Henley Centre Media Study struck me as telling us more about ourselves than the report's discovery that an underclass is addicted to television, but that higher income groups find it dull. On the day we published that last week, George Walden observed on another page that people had become thoroughly bored by Maastricht.

In the last half century, ways of telling people what is going on in the world have multiplied. Over the same period British interest in what is happening in the world has declined. I travel a certain amount, but I have learned to stay silent on all overseas experiences. Tell them what is happening in Angola, and watch the stifled yawns. Show them the photo-

people lose interest in what is happening in the world, despise their political leaders, and switch off, authoritarians bestir themselves.

★ ★ ★

AT WESTMINSTER last week I fell to wondering how far density of population affects human behaviour. On Thursday night, with three-line whips in both Houses, there must have been 1,000 parliamentarians in the Palace, and at least another 1,000 members of the public and staff.

I found it claustrophobic and could well envisage one man socking another on the jaw. When the troubles in Northern Ireland began in 1969, someone told me that the exceptionally high population density of Belfast made bad much worse. I believe it.

A big division in both

Battalion, as observer, all of them serving in the Air Corps, took off with other aircraft to bomb the city of Lubiana.

"At the altitude of the Selva di Ternova, the aforesaid Caproni was attacked by two enemy fighters with brisk machinegun fire. In an instant, Captain Salomone was wounded in the head. Lieutenant Colonel Barbieri, while preparing to fire his machinegun at the adversary, and the other pilot, Captain Bailo, who had fired rifle shots, fell gloriously and mortally wounded.

"Captain Salomone, sole survivor, had at least six times heard the crackle of machinegun fire and twice seen an enemy Fokker monoplane cross ahead, each time above but close enough to see the gestures of its pilot demanding surrender. Bearing in mind the sacred duty incumbent on him, and after having painfully found his bearings, nearly blinded by the blood flowing from his wound, he began the sad return journey with the indomitable and heroic intention of taking back the aeroplane and the glorious bodies of his fallen comrades to our lines.

"One of the engines had been hit and was not running well. Captain Salomone managed to shift the body of his colleague Captain Bailo, who, with a sublime sense of fraternal love, had tried, before dying, to shield the surviving pilot with his own body to protect him from the enemy's bullets. So sublime was Captain Salomone's sense of honour and duty that, far from succumbing to the enemy's signals, with admirable coolness and energy, with rare skill and the greatest valour, he continued dauntlessly on his flight to his homeland, committed to which, he brought back the aeroplane entrusted to him and the glorious remains of its brave sons, after having flown low over the enemy lines, where fortunately the anti-aircraft fire was in vain!"

The King bestowed on Salomone the first Gold Medal for Military Valour awarded to the Air Corps.

In January, d'Annunzio had begun to fly with Lieutenant Luigi Bologna, a naval pilot. On the 16th of that month they had taken off in a seaplane on a reconnaissance to Trieste, but a defect in the carburettor had forced them to put down at Grado. Perhaps because of mirror effect, or else dazzle, the pilot had misjudged his height above the water, "landed" too high, and the aircraft had dropped heavily onto the surface. The pilot got away with a few scratches. The observer hit his cheekbone on the side of the cockpit and sustained what seemed to be a superficial injury.

For the time being, d'Annunzio complained of great pain but refused treatment because he did not want to cause his companion to feel a guilty responsibility, seeing that the incident was clearly the result of pilot error. He continued to fly on operations without rest but fatigue aggravated the condition of his damaged eye to the point where a detached retina was diagnosed. There had also been haemorrhage. A little more than a month after the accident, he was ordered complete rest in darkness, which he had to endure for seven months.

In the meanwhile, the Commanding Officer in Venice recommended him for a decoration for "having repeatedly given proof of boldness in the air and for disregard of danger". He was duly given the Silver Medal for Military Valour. He also capitalised on his affliction by writing a much-praised poem, "Notturno".

The Poet now, in addition to a military cloak and silver-knobbed ebony cane – all Italian officers carried a walking stick – sported a thespian black eyepatch or, when he wore glasses, a blacked-out right lens.

He resumed flying on 13th September and wished Luigi Bologna to continue as his pilot, "to give his grieved companion an encouraging demonstration of trust and esteem". And no small amount of publicity for the Poet at the same time.

The resounding acclamation of d'Annunzio's participation in operations by submarine, aboard surface vessels and in the air was so good for the morale of the fighting Services that, in September 1916, the Supreme Command appointed him liaison officer of the 45th Division. With this appointment he shared in the life of the infantry, staying in the most advanced command posts, taking part in offensive actions, writing effective reports for Headquarters. He was promoted to captain for his part in the battle at Veliki Kriback on 10th to 12th October 1916; and, in the following month, decorated with another Silver Medal.

In January 1917 the French Government awarded him the Croix de Guerre. He also returned to flying on reconnaissance and bombing operations.

But there were other Italian airmen doing their bit. The official history describes the difficulties with which the Italian Air Corps was beset, in addition to the small numbers and poor performance of its aeroplanes: "The winter was severe, hostile to the combatants in the air and in the trenches; avalanches, more deadly than artillery, hindered the use of roads in the Alpine valleys; the snow and the piercing cold tormented the garrisons on the high peaks; the rain made the infantry suffer in the trenches where they were confined. It transformed them into quagmires. It flooded encampments and dumps. *The fog, the icy and violent winds, the whirling air currents above the mountains, the fierce north-east gale on the Carso, the impenetrable mists and clouds, were tremendous, often insuperable, obstacles to flying.*"

The slow reconnaissance aeroplanes often met such strong contrary winds that they could scarcely make headway against them. Beyond the Isonzo they were easy targets for the anti-aircraft artillery, which was growing more accurate and intense all the time. They would not turn back until the observers had completed their task, "the purpose of the sortie accomplished with honour, a sacred duty to perform".

The urgent necessity to perfect the means of reconnaissance led to rapid progress in radio telegraphy and photography. Concurrently, artillery observation gradually became more precise with practice. By the summer, on a twenty-five-kilometre front held by four army corps, four artillery co-operation squadrons were operating. Each kept three machines constantly airborne, one for reconnaissance and the other two for directing fire. So much had been done during 1916 to enlarge and improve the Air Corps, that it was able to look forward to the spring weather of 1917 with optimism.

In 1916 a twenty-eight-year-old Canadian, Harold E. Hartney, who was known throughout his RFC career as "Yank" and would, within two years, become one of the United States of America's most valuable air commanders, first saw service in France.

As a student Hartney vacillated in the choice of a career. At Toronto University he first studied Arts, then Engineering for two years, before graduating in 1911. Thence he went to Saskatoon University to read Law. In 1913 he began to practise as a lawyer, in partnership with his brother. His hobbies were playing the cornet and rifle shooting. It was not illogical, therefore, for him to join the Militia in 1911 as a private in the 105th Saskatoon Fusiliers. He was soon commissioned Second Lieutenant.

His regiment did not leave Canada until May 1915. Drilling near Dover, he saw a Farman Longhorn land nearby and immediately became so interested in flying that he applied in July to transfer to the RFC. He was accepted on 21st October; just as his battalion embarked for the Front. Meanwhile his wife had arrived in England and he led a pleasant domestic life while learning to fly in Norfolk. On 13th December 1915, with four hours ten minutes solo, during which he flew Longhorns, Shorthorns and BE2s, he got his Royal Aero Club certificate.

He also got rheumatic fever and went to hospital in January 1916. While there he was visited by seven fellow pupils. When he reached France in June, only one of them was still alive: the Fokkers had accounted for the rest.

His admiration for the British was genuine and generous: "Only a superb sense of organisation, a united opinion and a proved psychological system could have produced such results" in developing the RFC so quickly to its high standard. He also said, years afterwards: "The one thing that struck me most forcibly, in contrast with conditions I met in the United States Army later on, was the absence of selfishness, politics and bluff."

When Hartney left hospital he spent some time at an advanced flying school. His instructor there, when posted to France, had an embarrassing experience. In those days of crude dead-reckoning navigation, erratic compasses and inadequate meteorological information, particularly about winds, errors in direction finding were not uncommon. Added to these was this pilot's unfamiliarity with the terrain and consequent ignorance of landmarks. Entrusted with the delivery of the latest FE2d, with the new 275-h.p. Rolls-Royce engine, he landed on the German aerodrome at Lille and spent the rest of the war in a prison camp; while the enemy derived valuable information from their prize.

On 15th June 1916 Hartney sailed for France with nineteen hours solo flying in his logbook. He spent only twenty minutes at the Pilots' Pool in St Omer before continuing to his squadron, No. 20. The squadron was replacing its 160-h.p. Beardmore-engined FE2bs by FE2ds. As well as a more powerful engine, these had three machineguns: one intended for the pilot but accessible to the observer. There were six aircraft, eight pilots and eight observers on each of the three flights, all officers. They slept in wooden-hutted dormitories and each flight had its own mess. He and the other new arrival were "greeted with great courtesy but with a peculiar kindly derision", which he found was the custom with "Huns", new untried arrivals: the same derisive term with which the British referred to the enemy.

In a flight, he says "was an atmosphere something like college fraternities", and the whole idea of a flight in a British air squadron was "to have the officers live closely together and to associate closely at all times and to instill a sort of single minded unity in the entire group, on the ground and in the air, with a smooth chain of responsibility from the top down to the newest officer." So impressed was he with this principle, that he introduced it to the US Air Corps.

He praised his squadron commander: "Major Malcolm, as fine a soldier as I ever met, a six-footer from the Regular Army and every inch a man." When Hartney and the other new pilot came in, the CO promptly rose to greet them. He had other qualities than courtesy and manliness, as Hartney was about to discover.

The second new arrival was an American, who on the train had confided to Hartney that it was madness to send them to France "to fly in old untried kites", with only nineteen hours solo. If he was required to do this, he declared, he was going to crash deliberately on his first flight. To Hartney this sounded pusillanimous, but his companion had the courage to cut short Major Malcolm's greeting, with: "Major, I'm sure glad to meet you, sir, but I'm afraid I'm in the wrong church. You know, I'm a scout pilot."

The Major rewarded this frankness by ordering him to leave the squadron at once. Hartney's thought on this has a blurred sort of logic about it and just a touch of the self-righteous: "This was extremely embarrassing to me but secretly I rejoiced over the fate of the youth who had threatened to crack up his first machine, when planes were so scarce and, to me, so wonderful."

In contrast, Malcolm welcomed him warmly. "You have nineteen hours solo, I see, but hours mean nothing." This statement alone would have disquieted most pilots. "All you'll need are guts and loyalty." The guts, presumably, so that they could be spilled by a burst from a Fokker's Spandau the first time he showed his nose over the enemy lines with a mere nineteen hours' experience, and loyalty not to have any hard feelings in the next world towards his CO or the RFC's inadequate training programme.

Hartney did harbour one adverse criticism: the airfield was not flat, but saucer-shaped, small, with woods on two sides and a small lake at one end. On his first take-off, the Beardmore engine cut out and the Fee's nose dropped. He risked turning back and was lucky to glide in without stalling and killing himself and the observer. A few minutes later he took off again. The following morning he was sent, with an escorting aircraft, to look at the Front. After several more familiarisation flights he flew on his first operation on 30th June. To his astonishment he was allocated a brand new Fee with a Rolls-Royce engine. "... the proud possessor of one of the finest two-seaters on the whole Allied front ... my morale went sky-high ... I, the little mouse, began to yell 'Bring on them cats!'"

Thirty bombers were to attack balloon sites. Three fighter squadrons were each to provide ten aircraft so that every bomber would have its own protector. The fighters landed at the bomber aerodrome to refuel and have lunch. As the first bomber taxied for take-off across the bumpy ground, one of its phosphorus bombs detonated with a vivid flash and a loud noise. A moment later there was another sheet of blinding light and a loud explosion from the second bomber. A roaring conflagration spread. Exploding ammunition and flaming pieces of aeroplane slashed through the boiling smoke like rockets going off in every direction. Hartney rushed to move his own aeroplane to safety. Two hangars and five machines, including one of 20 Squadron's, had been destroyed. The remainder took off and carried on with the operation.

The next day the Battle of the Somme began and he was on dawn patrol with AM1 Stanley as gunner, with four other machines. His position was at the left rear – "outside left" – of the V. At 13,000 feet, as the flight began a left turn, he saw two enemy aircraft diving on Callender, who was flying "outside right". He fired a warning Vérey cartridge. The enemy streaked past, firing, then climbed to come in again. Hartney turned into them. One Fokker broke away. Callender did an Immelmann turn onto the other's tail. After some skirmishing he shot it down in flames. Hartney had lost 4500 feet and was under attack. Bullets crackled past his head. Stanley signed that both his guns were jammed. Hartney did a rough turning sideslip to the left, then another to the right. He couldn't see the enemy. He dived to gain speed and pulled up steeply. The controls went mushy and he kicked on hard right rudder to come heavily down and round in a stalled turn, a heavy-handed Immelmann. The Fokker was dead ahead and slightly below. Stanley had corrected the stoppages and was shooting: tracer, ball, incendiary and armour-piercing in sequence. Bullets riddled the Fokker and its pilot. The fuel tank burst into flames. Hartney's imagery becomes unexpectedly grisly for so kindly and prosaic a man: "I once saw a kitten, its head run over by an automobile, jump up in the air, then dive to the ground and roll over and over. Our enemy was doing just that. In addition, he was afire."

They were far behind enemy lines. Hartney turned for home. Stanley began to make agitated signals which meant that enemy aircraft were ahead and astern. Hartney saw two more Fokkers. Twisting, climbing, skidding, the three aeroplanes milled around dodging each other and trying to get their sights on a target. One Fokker was cautious and kept breaking off. The other was bolder, crossed ahead of the Fee, and Stanley shot it down. The Fee's engine was misfiring and Hartney had to make a forced landing close behind the British front line.

Later that day Major Malcolm was killed when his engine failed on take-off and he stalled in. He was replaced by Mansfield, whom Hartney said was "The typical high type of regular British Army officer, the same calibre as Malcolm, just as understanding, fair and helpful to budding young pilots."

On 21st July he was on the periphery of a tragedy of the kind that recur to one's mind in sombre moments for the rest of life. A sore throat and temperature kept him in bed instead of going on patrol. He listened to five aircraft of his flight, B, take off, and worried about the thickening fog. At lunch time the patrol had not returned. Nor did they return that afternoon. They had flown into a hill and all ten pilots and gunners were killed.

Early in September his face was so badly frostbitten at 14,000 feet that he was sent on a few days' leave in St Omer. On the 10th he went to England on leave. Three hours after rejoining his wife and baby daughter he received a telegram recalling him. Early next morning he ferried a Fee from Farnborough to his squadron and found that he was now deputy flight commander. On 8th December, with his statutory six months at the Front almost completed, he was offered the choice between joining a Home Defence squadron in Britain, or ten days' leave from 10th December and command of a flight on the next vacancy. He went on leave and returned on Christmas Day 1916 as commander of A Flight, with the rank of captain. As a result of casualties, he was already fourth in seniority on the squadron.

Nineteen-sixteen also saw the graduation into prominence of a strong personality who has been called "the man who taught the world to fly", on account of his revolutionary ideas about the training of embryo pilots. Robert Smith Barry was an Irish Old Etonian of rich landed family. When he joined the first course at the Central Flying School on 10th August 1912, he was a second lieutenant in the RFC Special Reserve, and an instructor at the Bristol Flying School.

In August 1914 he crossed the Channel as a member of No. 5 Squadron, crashed a couple of months later and was sent to hospital in England. The letter his squadron commander, Major J. F. A. Higgins, wrote him there is a model of the sort of letter every crashed pilot would be gratified to have from his CO. "My dear Smith Barry, I am awfully sorry about your smash but hope you will be alright in a few weeks and that in the meantime your breakages will cause you as little pain as possible. I am sending on the rest of your kit. Let Rabagliati [a squadron comrade] know if there is anything else we can do. Don't worry about the smash. I am quite sure it was not your fault. These things are bound to happen sometimes. It is lucky that it was no worse." Those were the days when a squadron CO could give a reassuring exoneration from blame, unfettered by the complications of a Court of Enquiry into the cause of an accident.

No doubt Smith Barry thought his injuries bad enough: he was still hobbling on sticks in the spring of 1915 when instructing at Northolt. Later that year he was instructing by day and flying anti-Zeppelin patrols by night. In November he was promoted to captain. In April 1916 he was posted as a flight commander to 60 Squadron, commanded by Major "Ferdie" Waldron, a fellow Etonian; who had chosen three of the same ilk to command his flights.

On 25th May the squadron went to France, equipped with Morane parasol biplanes. Next month six of these were replaced by the Morane Bullet, an aeroplane of dubious quality. Waldron was shot down and killed on 3rd July and command passed to Smith Barry. With his majority, the charming and popular Irishman was able to display his already notorious eccentricity regardless of criticism; not that he had ever taken much notice of higher authority. It had always been his pleasure to frighten his passengers into fits by subjecting them to extravagant aerobatics. Now, when an enemy single-seater mistakenly landed on 60's airfield in thick mist, he went beyond the bounds of the hospitality shown by both sides to captured fellow aircrew, and kept the bewildered pilot as a mess guest for a whole week.

He detested Trenchard, whom he condemned as a butcher. When pilots arrived on the squadron with as little as seven hours' solo, he refused to send them on patrol until he had brought them up to a standard which gave them some chance of surviving longer than the three weeks that were the average at the Front whenever the enemy had air dominance. Trenchard always insisted that he looked on the RFC as a family and tried to consider aircrews' feelings. That was why he had dead men replaced immediately: so that the others would have no time to brood over missing faces, but be more concerned with getting to know new ones. His family feeling did not deter him, when he sent three volunteers to attack balloons, from saying: "Good luck. But remember, it is far more important to get those balloons than to fail and come back." Using Le Prieur rockets, which had a 200-yard range, three between the wings on either side, they shot down all the balloons without loss.

From 1st July to 27th November 1916 on the Somme, the RFC lost 308 pilots, 191 observers and 782 aircraft. It destroyed 164 and drove down 205. There were many who felt as Smith Barry did.

He was also outspoken in his criticism of training methods and in November wrote a paper suggesting improvements. Trenchard neatly hoisted him with his own petard and had him posted to command No. 1 Reserve Squadron. Smith Barry was dined out by 60 Squadron on Christmas Eve 1916. Immediately on arrival at Gosport he set about ridding the training squadron of its antediluvian Shorthorns and Longhorns and forming one flight of Avros, another of BE2cs and a third of Moranes and Bristol Scouts.

The last word on this turbulent year of 1916 belongs to Henderson, who, at the start of 1917, summed it up in a memorandum to the Deputy Chief of the Imperial General Staff.

"The success of the Royal Flying Corps in the Field is not to be measured by the relative casualties to our and to the German aeroplanes; although this method of calculation would show a distinct balance in our favour, it does not bring out the great superiority which has been manifested by the Royal Flying Corps throughout the past year. The work of our Air Service is work done for the Army, and it is only by the success of that work – which nowadays is almost the same thing as the success of the Army – that the value of the Air Service should be measured. It would be easy enough to reduce the casualty list if we would consent to limit reconnaissance, to limit Artillery observation, to give up bombing, to concentrate all our available strength in fighting aeroplanes for the purpose of killing Germans in the air. No doubt our casualties would then be still fewer, and the German casualties still greater. For long periods at a time this policy has been adopted by the Germans with disastrous results to their operations on the ground. Throughout the Somme offensive last year, almost no German aeroplanes were seen, except fighting aeroplanes and they mostly operated ten miles behind the trench lines, and all through this period the German communiqués boasted of the success of their Air Service, and tried to measure that success by the number of British and French aeroplanes they had brought down; but the German Army was not to be taken in by those communiqués, the German Flying Corps was completely discredited in its own country, with the result that it has been completely reorganised, and is now making an effort to meet us on our own lines, but even now the work which is done for the German Army by its aeroplanes is very small compared to the value which our Army gets out of the Royal Flying Corps.

"Our casualties are heavy, but we must beat the Germans in the air, and the Flying Corps must carry on the work of the Army. Last year we beat the Germans in the air as soon as we had a sufficient spell of continuous good weather, and, in spite of the improved efficiency of the German Air Service, there is no reason why we should not beat them again this year. But there is no such thing as 'command of the air' at present, all that can be done is to press the air fighting further back behind the German lines, and try to keep the battlefield clear.

"With regard to the relative efficiency of British and German aeroplanes, a great many false comparisons have been made, and a great many discouraging statements made. It is most undesirable to quote in public either the specific types or the number of aeroplanes of different kinds in our possession, such information would be of the greatest value to our enemies. The facts at present are that in design and in improvement of design, we are now ahead of the Germans, but owing to the extraordinary rate of increase in the Royal Flying Corps, we are still behind the enemy in the matter of supply, and particularly in the matter of the replacement of types which are becoming outclassed. The circumstance is that our best machines, which are daily increasing in number, are better than the German best, but a certain number of ours are becoming outclassed by the average German machine. The necessary examination having, for the time being, been completed, our whole energy is now given to replacement, which is proceeding at a rapid rate.

"The continual attacks on the Air Services which are made in the Press and in Parliament, do an infinite amount of harm. They give information to the enemy, they discourage the officers and men, and they have a serious effect on the discipline of the Corps. No other branch of the Service, either Naval or Military, is exposed to this continual criticism from those who profess to be experts, and yet have no knowledge of the conditions at the Front."

He is loyal to Trenchard's policy in France, he puts air operations in perspective, and gives all categories of aircrew their due.

CHAPTER 13

———◆———

1917. The Climacteric

Not only was 1917 the climacteric, the period in which the greatest changes took place, but it was also the vintage year. During these twelve months all but a few of the most famous pilots reached their peak. The most conspicuous exceptions were the outstanding Americans, who did not feature until the next year.

So far, among those who were to shine in the US Air Service, only Hartney – who was still a Canadian citizen – and some of the Escadrille Lafayette had distinguished themselves. Lufbery was the sole member of the escadrille who was destined to figure among America's highest-scoring pilots. Of its original seven Americans, Chapman and Rockwell had been killed in action before the new year dawned and Prince had died from injuries received in a crash when making a night landing. At the Armistice, Hall and Thaw were the only survivors.

The word "climax" is commonly misused to mean "the highest point". It is in fact a rhetorical figure in which the sense rises gradually in a series of images, each exceeding the other. In that, the correct, sense, 1917 qualified also as the year of climax. It was the year in which aviation advanced from the shadow of its beginnings into the effulgence of its future; from the era of the Wright Brothers to that of R. J. Mitchell, who fourteen years hence would design the Schneider-Trophy-winning Supermarine S6B from which he derived the Spitfire.

In the early 1800s Sydney Smith, clergyman and journalist, unwittingly summarised the practical philosophy of fighting airmen a century later in their attitude to life: "Take short views, hope for the best, and trust in God." The first two articles in this creed were forced upon them by the circumstances of war. The third, they mostly amended to "trust in yourself". It was by experience combined with natural ability that pilots in the two great wars might achieve longevity. Observers and gunners had to trust in their pilots.

The physical and mental ravages of combat flying, whether on reconnaissance, artillery co-operation or fighter patrols, can be seen in contrasting photographs taken of airmen at the start and towards the end of their operational tours. By the time that Louis Strange and Hawker were transferred to the Home Establishment after their first eleven months at the Front, they looked weary, haunted, stunned and brittle. Boelcke, fleshy-faced and youthful in 1914, died hollow-eyed and gaunt. Guynemer was always delicate-looking. By the time the Somme battles were done with he had dark circles under his eyes and looked as though he might collapse from exhaustion. Fonck, Ball, Immelmann, all bore the same signs of over-stressed nerves, lack of sleep, the erosion of long hours in arctic cold with the brain bemused, heart and lungs labouring, from lack of oxygen.

Parallel with the improving performance of aeroplanes, the weapons with which they were armed, and the artillery that threatened them from the ground, amidst all the overheated and often melodramatic politics of war, another concern than those that dealt with destruction was stirring. Typically, Henderson was at its core.

On 6th February 1917 the Royal Flying Corps Hospital was opened. This was the culmination of a plan that had begun in early 1915, when a small hospital for RFC officers was started by Dr Atkin Swan and Mrs Paynter. At first it consisted of a few beds in a private London nursing home. Later they rented a small house which would accommodate twelve patients. Then Lady Tredegar offered her house, 37 Bryanston Square, with a large contribution towards equipment and upkeep. By May 1916, 350 cases had been treated there: wounded from the Front, victims of accidents at home, and others whose health had been impaired by continuous flying. Many dangerous and delicate operations had been performed, by means of which lives were saved and officers made fit to resume their duties. In addition, 1100 out-patients had been treated, almost all by Dr Atkin Swan. The whole medical staff and other workers gave their services voluntarily.

The existing accommodation had become inadequate and a thirty-one-bed extension was being opened at 82 Eaton Square. The increased expenditure meant that the hospital could no longer exist on private contributions. Among the Henderson papers is an appeal he evidently sent out, but no indication of its circulation.

"I am sure, if those who read this could only see the patients, whose one wish is to get back to their work, if they could know their indomitable spirit, and hear some of their stories of deeds done in the air, with utter simplicity which makes one proud to be of their race, they would give to their utmost to help these boys to regain the use of their broken limbs and their nerves, which are often shattered for the time being by all they have been through."

At last here was official recognition of battle fatigue and sympathy for it, not revilement. Medical treatment for suffering minds and spirits was proffered instead of a brusque order to undertake even more than usually trying duties in the air or to serve for a spell in the trenches.

The War Office and Admiralty had ideas that differed from Henderson's and were advocating a joint hospital in which RFC and Naval officers would be treated. The latter would include the RNAS; but as this was a much smaller Service than the RFC, naval Medical Officers had comparatively little experience of treating patients suffering from psychological illnesses arising from operational flying.

On 14th June 1917, Henderson wrote to the Adjutant General. "I beg that you will submit this question to the Secretary of State for reconsideration. I have only now seen this paper for the first time, and there are a good many points which have not been considered at all.

"1. Present experience of the peculiar ailments to which flying men are subjected is too valuable to be wasted. A great deal of knowledge is now possessed by the Standing Medical Board of the RFC and the staff of RFC Hospitals. Far more attention is being given by the Military than the Naval authorities. If officers of the RFC were to be relegated to a Naval Hospital, the whole of this knowledge and experience would be lost. There have been, ever since the beginning of the war, Medical Officers giving special attention to the RFC in France and they have accumulated the most knowledge.

"2. If the Army Medical Authorities consider they are unable to look after RFC sick and wounded due to temporary difficulties with regard to medical personnel, the whole of the experience of the war will be lost to them and when the war is over the RFC will either have to be dependent on the Royal Navy for medical advice, or to subject themselves to medical experiments under the hands of Doctors to whom the whole subject will be absolutely new.

"3. In the Military Wing of the RFC alone there is an average sick list of something over 400 officers, including wounded. Through private enterprise, we are able to deal with a certain number in RFC Hospitals (51 beds) and 70 beds in Auxiliary Hospitals in the country. Therefore there are still some 300 cases which should be concentrated. I cannot see that they would require any larger staff when concentrated than while distributed among Military Hospitals.

"The number will certainly not diminish and is quite large enough to justify a separate Military Hospital without taking into account any possible Naval cases.

"With regard to accommodation, I do not think it would be possible to work the present RFC Hospitals under Naval control or in conjunction with the Naval Hospital. There would also be many difficulties in combining the activities of the present Standing Medical Board under Naval administration.

"4. After considerable delay and difficulty, we now have working a system by which the Standing Medical Board is able to keep records, more or less complete, of flying officers whose sickness or injuries cause any element of doubt about fitness for future flying.

"These cases are peculiar, and require watching, and although there are great difficulties under the present system, where a large number of officers are not in special Hospitals, yet the whole organisation is Military, it is possible to do something towards this end. Civil staff of RFC Hospitals give great assistance in this respect but as they give their services voluntarily, it is necessary to establish and maintain personal friendly relations with them, and I am extremely doubtful of the attitude they would adopt if the scheme were put through.

"5. I see no particular advantage in a joint Hospital because I do not think we are likely to obtain a single hospital large enough to contain all the Military cases alone. If, however, a joint hospital is insisted upon, it seems to me absolutely essential that it should be under Military control, if only for the reason that the whole of the organisation for treatment of and research into the peculiar ailments due to the Air have been carried out by the Army. As far as I have been able to ascertain, the Navy Medical Service has been very little acquainted with these problems."

There was no bigotry about Henderson's stance. He was not moved by jealousy of the Royal Navy. He knew what was best for the Royal Flying Corps and had taken part in the efforts that had been made privately to provide it. As well as the depth of compassion to be concerned about the plight of those serving under him, he had the breadth of vision to be concerned about the effect that neglect of their needs would have on the efficiency of the Service.

He submitted his plea to the Chief of the Imperial General Staff. "One of the most important needs for the Air Service is a specialised Medical Service, but in this case, as in a good many other matters, the urgent necessity has only become apparent during the last year. It has been found by experience that flying men are subject to many peculiar physical disabilities, and research into methods of prevention and cure of these disabilities has advanced very rapidly. So long, however, as the Flying Services remain merely minor appendages of the Army and Navy, it is very difficult to provide for the study of these medical matters and for the special treatment of the patients.

"Knowledge on the subject is limited almost exclusively to a few Medical Officers and civilian Doctors who have had considerable numbers of flying men continuously under their care, and it is necessary at once to take special measures to provide separate accommodation for these cases and separate staff to concentrate on the problems which arise. The mere selection of candidates for the Flying Services has become a highly specialised business and the standard of fitness required for flying is diverging from the standard for ordinary Naval or Military officers.

"Also there are many minor physical reasons which render a man quite unfitted for flying, yet may, if taken in time, be removed or cured.

"More important are the many precautions which have to be taken to enable pilots to maintain their fitness under the severe conditions of modern air warfare.

"It is not too much to say that what has been called the strain of flying, and has in the past been put down entirely to nervous causes, is very frequently the result of physical conditions which can be ameliorated. It is probable that the advance of medical science may tend to prolong considerably what is called the life of a pilot, i.e. the period for which he can continue active flying.

"It is only by means of a Medical Service devoted entirely to the study of aeronautical problems that real progress can be ensured."

Henderson's interest in, advocacy of and ultimate success with Aviation Medicine benefited the Royal Flying Corps, and ultimately the Royal Air Force, as greatly as any of Trenchard's achievements in other spheres of crucial importance to the existing and future Service

Saving lives and sanity, healing wounds, repairing broken limbs, and the internal injuries common when aeroplanes crashed, was all very well; but analysis of the ways in which the total number of those killed at the Front had met their deaths indicates that one-third could have been saved by the provision of parachutes.

Balloon observers on both sides used parachutes similar to the one that Pégoud had demonstrated at Hendon in 1913. They were contained in a bag attached outside the basket or aircraft and connected to the man by a long cord, which tugged the canopy open when he jumped. They could have been used by Service aircrews, although many would have been riddled with bullets or consumed by flames and rendered useless. Designers were working on one, to be worn by pilots and observers, which was compact and less at risk from damage.

Whatever form such a life-saving piece of equipment might have taken, it seems that it was opposed by the senior officers who sat secure from danger at staff desks. The reasons put forward for prohibiting parachutes was, at best, a parody of the code of Victorian "manliness": balloon observers could resort to a parachute without shame, because they were defenceless (apart from the concentration of machineguns and anti-aircraft cannon sited specially to protect them, which were evidently ignored). Aeroplanes, on the other hand, carried weapons: their crews were therefore not entitled to a means of escape. They must rely on their flying skill and good shooting to save their lives. The Air Board advocated, instead of parachutes, armour plating to protect aircraft and their occupants.

The Air Board condemned itself. "It is the opinion of the Board that the present form of parachute is not suitable for use in aeroplanes and should only be used by balloon observers.

"It is also the opinion of the Board that the presence of such an apparatus might impair the fighting spirit of pilots and cause them to abandon machines which might otherwise be capable of returning to base for repair."

In fact, the Board harboured the insulting suspicion that pilots would be tempted to abandon combat and jump for their lives in moments of extreme disadvantage instead of pressing on and fighting to the death. If a naval captain traditionally went down with his ship rather than take to a lifeboat or put on a life jacket and swim, why shouldn't a pilot or observer display the same phlegm?

Those who should have provided for the saving of life by parachute were guilty of ignorance and disregard for human life; which they evidently rated less important than the salvage of aircraft.

As early as 1908 a parachute worn by the jumper and opened by a ripcord he operated had been invented in the United States and often displayed during the next four years. This could easily have been copied or bought by all the world's air forces for their aeroplanes. Nobody gave the matter any attention.

In 1914 a civilian, E. R. Calthrop, had perfected a parachute that, although it had to be attached to the fuselage, was smaller and more efficient than the Spencer type fitted to balloons. He gave it the unfortunate name "Guardian Angel", whose connotation of a convenient means of avoiding danger prejudiced the Admiralty and War Office against it. He seems also to have been less than diplomatic in his approach to these pompous authorities. The RNAS and RFC refused to attend a demonstration.

In 1915 the Superintendent of the Royal Aircraft Factory wished to make experiments with Calthrop's parachute. He informed the Directorate of Military Aviation that he had fitted one to an aeroplane and was ready to carry out tests with a dummy. He asked if the Directorate wished him to proceed.

Henderson himself replied and his words are astonishing after his general humanitarian concern and the many occasions on which he had shown high intelligence and foresight. He wrote back: "Certainly not." But then, this was the man who had compared loss of life in war with the breaking of eggs for omelettes. So perhaps there was not so wide a gulf between his code and conscience and Trenchard's after all.

Although Calthrop persisted in trying to press his invention on the authorities, that was the end of the matter.

Trenchard emerges from the controversy with credit. In 1917 he suggested that trials should be made in France. This was turned down: so he asked for twenty of Calthrop's parachutes for dropping agents behind enemy lines; and got them.

A ludicrous intervention was made by a sixty-nine-year-old member of the Air Board, Lord Sydenham, who had retired from the Army in 1896 and had never set foot in an aeroplane. He queried whether a pilot would have time to think of taking to his parachute in the throes of coping with an emergency. This absurdity was endorsed by General Groves, whose written comment was that a stricken aeroplane usually fell so fast that a pilot had no time to think of abandoning it. Groves had never flown either.

The improvement of parachutes continued, but none was issued to any of the Allied air forces during this war. A few Germans were seen to use them in its last few months.

The Allied Commanders-in-Chief tried to alleviate the general gloom that had prevailed since the Somme by planning a Spring Offensive which would enable the 3,900,000 British, Commonwealth, French and Belgian troops to win substantial ground from the 2,500,000 Germans who opposed them.

In the air, Germany had thirty-three Jagdstaffeln and was forming more. The RFC and Aviation Militaire were recruiting vigorously and expanding. Trenchard informed Haig that he would need twelve long-range bomber squadrons and fifty-six fighters and Army co-operation.

The Albatros DIII and Halberstadt DII were better than any British fighter except the Sopwith Pup. This had first been issued to the RNAS in the last quarter of 1916. Armed with a Vickers gun synchronised with the propeller, a delight to handle, responsive, swift and free of vice, it climbed to 10,000 feet in under twelve and a half minutes and its speed at that height was 104 m.p.h. It instantly proved a success in action.

In October 1916 Trenchard had asked the Admiralty for help: eighteen fighters from those based at Dunkirk. His urgent plea was granted. A new squadron, Naval Eight, was formed with one flight of Pups, one of Nieuport 11s and one of two-seater Sopwith 1½-strutters. In November the two-seaters were replaced by Pups, which, in December, also replaced the Nieuports. By the end of the month the Pups had shot down twenty of the enemy and the Nieuports two. The squadron had lost two pilots. Early in 1917 it returned to its own Service.

There was now also one Pup squadron in the RFC, No. 54. The Germans admitted the Pup's speed, manoeuvrability and fighting qualities. On 4th January Flight Lieutenant A. S. Todd of Naval Eight fought three Albatroses led by Richthofen, who reported: "A Sopwith one-seater attacked us and we saw at once that the enemy aircraft was superior to ours. Only because we were three against one we detected its weak points. I managed to get behind him and shot him down."

The current Albatros and Halberstadt were superior to the French fighters except the Spad, which had begun to reach the Front by the end of 1916. Spad was the acronym for La Société pour l'Aviation et ses Dérives. The S7 of this marque was the fastest aeroplane at the Front and carried one Vickers gun firing through the propeller.

The Italian Air Corps was also increasing in numbers and improving the quality of its equipment. The aviation industry was growing. Static warfare made it possible to improve living conditions and technical facilities on the aerodromes. Photographic reconnaissance was producing excellent results and dominated winter activities. Preparations were going forward for the Tenth Battle of the Isonzo, in the spring.

McCudden, who had, as an observer, been promoted to flight sergeant in January 1916, was sent home that month for pilot training; at last. On 8th July he was back in France as a flight sergeant pilot on No. 20 Squadron. In August he had been transferred to 29 Squadron to fly DH2s and had made his first kill as pilot on 6th September.

On New Year's Day 1917 he was commissioned as a second lieutenant. He scored his second victory on 26th January; and, by the time he was posted to England as an instructor, on 23rd February, he had shot down three more enemy aeroplanes.

Mannock, whose career as a fighter pilot ran parallel with McCudden's, had applied for a transfer from the Royal Army Medical Corps to the Royal Engineers Signals Section as an officer cadet. In March 1916 he was a sergeant major in his new corps; and still burning with class hatred. He told his civilian friends that he had to work extra hard to compete with the better-educated candidates, loathed the conversation of the ruling class and hated them almost as much as he did Germans. While on leave he met a friend who was now an RFC pilot and this, together with the newspaper accounts of Ball's exploits, attracted him to the idea of flying. As soon as he was commissioned, in June 1916, he had applied for the RFC and had begun his ground training in August. On 28th November he had passed the Royal Aero Club test. On 5th December he had joined No. 19 Training Squadron, then began a course at the Hythe School of Gunnery on 1st February 1917.

On 15th February he was posted to No. 10 Reserve Training Squadron to learn to fly the DH2 and FE8. Here he met McCudden, who made a strong favourable impression on him; and who, in return, spoke of him as having great natural aptitude.

In January, Sholto Douglas, newly returned to France in command of 43 Squadron, found snow on the aerodrome at Treizennes, where 40 Squadron was also based. One of his flight commanders had stalled on take-off from England and was killed with his mechanic. Another, with two broken legs, had to walk on sticks and be lifted into his cockpit. Morale, already affected by the Albatros and Halberstadt, was not high. Most observers were still not being given any training in Britain and had to be taught their duties on the squadron. The 1½-strutter was fragile and broke up if roughly handled. To show what could be done with it if treated properly, Douglas took his Recording Officer, Purdey, up and did thirteen loops. It was not until he landed that he found that Purdey had not done up his safety strap. Centrifugal force and, no doubt, a frenzied grip on the cockpit coaming, had kept him in his seat.

For defence against enemy fighters Douglas made his formations fly with less than a length separation between aeroplanes and fight in formation instead of breaking. From the first few sorties he learned that offensive patrols were going to have to fight all the way out and back. As soon as they crossed the lines they could see German fighters taking up favourable positions to attack them. Casualties in February and March were so heavy that someone was lost almost daily. These losses were caused not only by the quality of the enemy machines and pilots but also because pilots were still arriving at the Front with far too little training and were often shot down within a few days. Trenchard admitted that training was inadequate but did not relax his demands for expansion. He also opposed escorts for bombers, which he maintained should be able to fight their own way out of trouble. Despite this, many raids were escorted because Brigade and Wing Commanders were in closer touch with reality than he was.

The RFC was on the threshold of its period of greatest adversity. On 9th March 1917 Richthofen led an attack on an offensive patrol of eight 40 Squadron FE8s in which four of them were shot down. Four others were damaged. On the 24th two Sopwith 1½-strutters of 70 Squadron on reconnaissance were shot down and the rest damaged. The task was repeated next day. The only survivor was one Sopwith that returned early with engine trouble. The other five perished. Thus fourteen officers were lost, dead or missing, and seven aeroplanes, in successive days and on the same mission. Total aircraft losses in March were 120, of which 61 fell behind enemy lines.

On 1st January 1917 Leutnant Carl Degelow, who was to be the last recipient of the Pour le Mérite, joined his first squadron, on the Western Front. He has been described as "Germany's Last Knight of the Air", but this is purely dithyrambic: he had no title. He seems to have been a decent enough sort who enjoyed consorting with captured British airmen.

At the age of nineteen he had joined the infantry in August 1914 and been commissioned in July 1915. He served first on the Western, then the Eastern Front, and returned to France in 1916. He volunteered to fly because he felt ashamed that he had lived in a dugout on a quiet sector of the line while there was heavy fighting at Verdun. In May 1916 he made his first dual flight. It was nearly his last, because of the unexpected effect of torque and a very strong wind. The Chief Instructor made him do twenty more. At the end of December 1916 he was posted to 216 Reconnaissance Unit.

His first photographic reconnaissance sortie also almost proved fatal. He was given an elderly Albatros CV, known as "the Furniture Wagon", which was much slower than the others. His two companion aircraft left him far astern across the French lines. First anti-aircraft fire and then five Nieuports did their best to kill him. In combat with the French fighters, his front gun jammed, his observer was wounded and the engine began to emit smoke. Resigned to being taken prisoner or burned to death, he switched off the engine and dived. Close above the treetops he switched on again and managed to reach his own side of the lines.

Soon after, he saw a Caudron spotting for French artillery and chased it. This was not his unit's function, but as he and his observer were both officers of the same rank, the latter could not exert authority. The Caudron opened fire. Instead of positioning the Albatros so that his observer could fire, Degelow went below it and began to shoot. Then his observer put in a few bursts. The Caudron crashed. Two months later they shot down another, in flames; and, a fortnight after that, had an inconclusive fight with a third, which they saw going down, smoking. The two definite victories were confirmed and pilot and observer were each credited with them. Degelow thought this illogical, as it implied four kills. The score was then adjusted to half a kill for each.

His Commanding Officer praised the two officers' aggressive spirit, but pointed out that shooting down the enemy was not what they were paid to do. Their aeroplane had suffered much damage and they had risked losing the valuable photographs they had taken. Leutnant Degelow showed his disgruntlement at being forbidden to attack enemy aeroplanes. His CO suggested that, as he plainly did not have the temperament for photo. recce., he had better transfer to a Jagdstaffel. In the meanwhile he paired him with an Oberleutnant observer who was a regular, as well as his superior, and who firmly restrained him.

At the end of July Degelow started a two-week course at the Kampfeinsitzerschule – Single-Seater Fighter School – at Valenciennes; whence he went on to Jasta 36: where we shall catch up with him.

He now rashly propounded two principles. One, "Immer ran auf meter, meter ... Ever closer, metre by metre," was trite. The other stretches credulity. He said that it was essential to get as close to an enemy aircraft as possible, which was already established by British, French, Italian, American and other German pilots and is indisputable. But when he added "even three or four metres", he was being boastful and foolish. To be so close to débris resulting from one's own gunfire, and, even more dangerous, flames and an exploding petrol tank, would be ridiculous rather than brave. His implication that it was his practice to approach to so short a range is a prime specimen of the type of indiscretion that the RFC used to, and the RAF still does, dismiss with derisive laughter and the condemnation "shooting a line"; or, more inelegantly, "bullshit".

Someone who more closely approached authentic knighthood was Ritter von Greim, "der Tank-Stösser ... The Tank-Buster". At the end of November 1916 he was still an observer, but at last about to be admitted to flying school. After his course he returned to Feld-fliegerabteilung 3, whose designation had been changed to Flieger Abteilung 46. For a while he continued reconnaissance flying, before his request to go onto fighters was granted. In April 1917 he joined Jagdstaffel 34, commanded by Oberleutnant Dostler. There were still many months to elapse before he began to earn his tank-busting fame.

*   *   *

Despite the extreme severity of that winter, a high level of air activity had been maintained on the Western Front. On the Italo–Austrian Front foul weather had greatly curtailed it. With the first signs of spring, in March, Allied casualties began to mount: the Germans were ascendant.

April brought a resurgence of reconnaissance, artillery observation and fighter patrolling in Italy. In France, it cast the threat of irreversible catastrophe over the RFC and Aviation Militaire. But Allied confidence and optimism were renewed when, on 6th April 1917, the United States of America declared war on Germany.

CHAPTER 14

——◆——

1917. Crisis

A crisis is a turning point and in 1917 there were two. The fourth month of that year has gone down in RFC history as "Bloody April". In the first week of April Henderson put the aerial situation at the Western Front, and policy regarding it, in perspective with a review and assessment.

"The increased number of casualties in the Field lately are due to several causes. In the first place, the retirement of the Germans over a large section of the front necessitated a great amount of long-distance reconnaissance and of photography. This is always dangerous work, and specially dangerous in this case because of the special efforts made by the Germans to stop it. I understand, however, that the information supplied by the Royal Flying Corps, as to the German movements and the German preparations in front of our Army, have been absolutely complete. This could be ascertained by reference to Sir Douglas Haig.

"Probably, in view of this retirement, the Germans had concentrated a very large proportion of their available forces in front of the British. There has not been nearly so much fighting in the French part of the line, and this may also be due to the fact that the French Air Service has hardly been pulling its weight of late. I was informed at General Headquarters that the information obtained by the French Air Service, with regard to their front, was very incomplete, so much so that a considerable portion of the German line in front of the French had to be photographed by the British Flying Corps.

"There is no doubt that the Germans have produced within the last few months, a considerable number of fast single-seater Scouts, of which the best is the Albatross [sic]. The aeroplanes which we have on our front which are equal to, or better than the Albatross Scout, are two French types – the Spad and the Nieuport – and the English Sopwith triplane: of these we have* squadrons in all. Next to them, and still able to hold their own, are the small Sopwiths and the Martinsyde squadrons. Our first-class two-seater machines capable of being used for offensive fighting, are the de Havilland 4 and the Bristol Fighter: there are at the moment 1 squadron of each. The FE2D, with the Rolls-Royce, is a two-seater Fighter, which will not be outclassed for some time: of these there are 3 squadrons. The machines principally used for reconnaissance are Sopwith $1\frac{1}{2}$ strutters: of these there are three squadrons. A squadron of SE5 single-seater Fighters, which is believed to be superior to any German machine, is due to leave England this week.

"The delay in producing larger numbers of these fighting machines is due almost entirely to the delays in engine production. We are only now beginning to get British made engines equal to those which the Germans had for the last eighteen months, with the exception of the Rolls-Royce engine, of which the supply has always been limited. The high powered British engines, however, have now reached the production stage, and the quantities delivered are expected to increase week by week, which will enable us to provide for the Expeditionary Force first-class fighting machines in good quantities.

"In addition to long reconnaissance, a very large amount of Artillery observation work is always going on, much more in our Army than in either the French or the German. This certainly adds to our casualty list without inflicting on the enemy proportionate losses in the air. It does, however, enable our Artillery to inflict much more serious losses on the German forces on the ground, and this must be taken into account in considering whether we get sufficient value for the casualties we suffer.

"With regard to the losses inflicted on the Germans, the announcements which are made in the official communiqués do not show their full extent; so much of the fighting takes place on the German side of the lines that very often there is no information whatever about the actions of our aeroplanes which are reported missing, but it is known that frequently in these unseen fights serious losses are inflicted on the Germans. The German casualties which are reported in our official communiqués are only those which are seen and vouched for by our Flying Corps in the course of their work, but from time to time fights have been witnessed from the ground in which both German and British

*Henderson omitted the figure, presumably for his staff to supply. The only Triplane squadrons were in the RNAS.

aeroplanes fell in German territory. But, considering even the accounts of fighting in the air, the losses on each side are not disproportionate, considering the different employment that is made of the air forces, that is to say that the German aeroplanes are merely employed in trying to bring down our aeroplanes, whereas ours are mainly employed in doing work required by the Army.

"It was noticeable last year that up to the beginning of June there was no marked superiority in the air on either side, and that the losses on each side appeared to be about equal. After that date, in the continuous good weather, our superiority became more and more marked, but our losses did not diminish to any great extent, for the reason that our superiority on the battlefield was only sustained by continuous fighting at a distance behind the German lines.

"If we would consent to adopt the same policy as the Germans, there is no doubt that our casualties in the air could be diminished. Hitherto, when the German has found himself inferior, he has given up reconnaissance entirely, and has confined himself to defensive fighting on his own ground, but if we were now to follow these tactics the effect on the Army generally would be most serious; we would be able to show an admirable balance sheet of casualties in the air, but the Germans would have information of our movements, and we would have none of their movements; they would have observation for their guns, and our gunners would be blind. Such a policy at this period would be disastrous. The casualties must be faced."

However much one admires Henderson, this clumsily phrased, execrably punctuated and often ungrammatical effusion compares poorly with the correct and usually elegant style in which senior French, Italian and German officers wrote.

The content has its errors too: neither the Martinsyde nor the FE2d was a match for the Albatros.

It would have been interesting if Henderson had written another report to the CIGS four weeks later to explain the calamity that his optimistic purview had not foreseen.

A massive attack by the British ground forces, the Battle of Arras, was planned to begin on the 9th April. As a prelude, the RFC opened an air offensive on the 4th. The intention was to drive the Luftstreitkräfte out of the battle area, so that contact patrol and artillery co-operation squadrons would be unmolested. In numbers, the British were greater. There were 41 squadrons in France, comprising 754 aircraft. Of these, 25 squadrons, numbering 365 serviceable aeroplanes out of a strength of 465, were on the First and Third Armies' front: opposed by 195 enemy aeroplanes, of which about half were fighters.

Both the Allies and the Germans had been learning since the year began that it was not numerical superiority that decided air battles: it was the performance of the aircraft. This was chasteningly demonstrated now. Between 4th and 8th April, seventy-five British machines fell. Nineteen lives were lost, thirteen aircrew were wounded and seventy-three went missing. At the same time hasty training of pilots, who were not fully competent when they arrived at the Front, resulted in the wrecking of fifty-six machines in accidents. These new pilots averaged only $17\frac{1}{2}$ flying hours, a mere ten solo, and had no experience of the types they were to fly on operations.

The statistics and Combat Reports make ugly reading. A typical disaster occurred on the 6th, when a whole formation of five DH2s of 57 Squadron on offensive patrol was shot down. A week later, when four RE8s ("Harry Tates": he was a famous comedian) were ludicrously supposed to be escorting two others, they met Richthofen's Jasta. He helped himself to one and his companions shot down the rest. Three days after that, four Albatroses shot down four out of six Nieuports. The enormity of sending the RE8 to the Front in late 1916 cannot be excused by a plea of necessity. It was designed on pre-1914 principles and was so slow and unmanoeuvrable that the observer was instructed not to stand and look over his pilot's head when about to land: because his added wind resistance would tip the flying – just about – machine out of the sky and kill or cripple them both.

The Harry Tate was Richthofen's 41st victory, which took him past Boelcke's record. Later the same day he added another, a BE2c. Both were as cheap as if a villain armed with a cosh had crept up on two old ladies in a dark alley and felled them.

The Germans had taken advantage of the four months' respite since the Somme battles to form and train new Jastas, to man them with their best pilots and to equip them with the finest fighter aircraft in the world. But these measures alone did not account for their high success rate. The British made it easier for them by continuing to fly creaking old BE2c and other obsolescent types. These, beating into the strong headwinds of a delayed spring – it snowed that Easter – were like sitting duck to a poacher with a double-barrelled twelve-bore. Moreover, when these travesties of what a contemporary aeroplane should have been were sent pottering off about their lawful occasions they had to be wastefully escorted by an equal number of FE2b or d so-called fighters and given a top cover of the same number of Pups: which did at least stand some chance against the Albatros DIII and Halberstadt; but needed to operate in greater strength to be fully effective. Whatever type or quantity of really good fighters the RFC might have had, the rate at which Trenchard was killing off pilots, including many of the most experienced, meant that there would not have been enough first-class men to fly them. Early in the year, he had forbidden squadron commanders to cross enemy lines. The best ones ignored the order as often as they could. In all, his policy of showing aggression at any cost did for roughly forty per cent of his pilots and observers. Not only offensive patrols and reconnaissance sorties went far behind enemy lines, but also long-range bombing raids.

To complement the British offensive, the French opened theirs on 16th April, on the Aisne. Commandant du Peuty imposed the same demands on l'Aviation Militaire. "Your task is to seek out, fight and destroy enemy aircraft," he directed; and "Victory in the air must precede victory on land."

They justified their policy by reasoning that the intrusion of their aircraft deep in enemy territory unsettled the populace, kept fighters tied up far from the Front, and occupied the attention of anti-aircraft batteries that could otherwise have moved up to the battle zone.

There was mutual esteem as well as collaboration. On 10th March du Peuty wrote to Trenchard:

"I do not know how to express to you all the admiration I feel, and the whole French Flying Service feels, for the British Flying Corps.

"These results have not only contributed to the great success of your armies, but in close co-operation with our own efforts they have relieved us of a large part of the German aviation.

"I hope to be able to teach what is left of the German aviation that the French intend to follow the same principles in the same manner. I should be grateful if you would let those under your orders know the admiration which the French Flying Service feels for them, as well as the feeling of comradeship they have for them."

He was short of the means to achieve his aspirations. Inefficient staff work had provided him with a force of only four Groupes, a total establishment of 200 aircraft. He had only 131 aeroplanes when the Battle of the Aisne began, and 151, the maximum, five days later. Among these the Spad S7 was competent to challenge the enemy fighters, but there were still too many Voisin and Farman pushers.

The Germans facing the French kept six standing patrols airborne: four at 6000 to 8000 feet and two at any height between 12,000 and 20,000 feet. The lower ones effectively turned away contact and artillery-spotting flights, while the high cover ignored the offensive patrols: which then merely wasted petrol, of which the French were suffering a dearth.

For the British fighters, trench strafing became increasingly frequent as co-operation with the artillery produced ever swifter results. The gunners were able so soon to get onto the targets – usually enemy batteries – given to them from the air, that the pilots were free to turn their attention to attacking the infantry.

The Germans had a two-seater, the AEG C IV, carrying a 200-pound bombload, with an armoured belly and armed with a Spandau and a Parabellum, which was excellent for this purpose. Two Staffeln of a new type, the Schlachtstaffel – battle squadron – were formed and equipped with these.

The British also had a new two-seater, for a totally different purpose, in which high hope was invested: the Bristol Fighter. This was a big strong machine over twenty-six feet long, with a wingspan of nearly forty feet. The crew sat closely back-to-back in a shared cockpit. There was a Vickers synchronised with the propeller for the pilot, and a Lewis, mounted on a Scarff ring for the observer to protect flanks and rear. The first series had a top speed of 111 m.p.h. at 10,000 feet, a height it could reach in 13 minutes.

The first Brisfit squadron, No. 48, arrived in France in late March, initially equipped with only six. Had there been time for the crews thoroughly to familiarise themselves with the aeroplane before taking it into action, its impact on the enemy would have been as startling and gratifying as the RFC expected. Tragically, instead of raising the Corps's morale and lowering the enemy's, the reverse happened. On 5th April – much too soon – a flight commander, Captain Leefe Robinson, who had won a VC for shooting down a Zeppelin near London, had to lead all six on their first patrol.

Five Albatros DIIIs, led by Richthofen, intercepted them.

The Brisfit pilots, following the standard two-seater routine, turned away from the enemy and took no evasive action, trying to give their observers a stable gun platform. Nobody had had time to find out that the excellently manoeuvrable Bristol Fighter could be thrown about like a single-seater; and that it was the pilot's heavy Vickers machinegun that should be used as the main weapon; with the movable and lighter Lewis as a means of defence, not attack.

As a consequence of this staid old-fashioned manoeuvre, Richthofen was able to shoot two Bristols down while his wing men took out two more. Of the two that got away, one was badly damaged. Leefe Robinson had gone down and was taken prisoner.

Interviewed by the press, Richthofen described the Bristol Fighter with contempt. This was no service to his comrades, who thenceforth attacked it over-confidently. The British pilots and gunners quickly learned how to fight their aircraft effectively. The manufacturers progressively gave it a more powerful engine, until its speed reached 125 m.p.h. In a very short time it became enemy doctrine never to attack three or more Brisfits, even when outnumbering them two or three to one. In order to tempt them to fight, Bristol Fighters used to go out in pairs and singly; but the Germans remained reluctant. This remarkable aeroplane, perhaps the best and most versatile of the war, remained in RAF service until 1932.

There was another new fighter of which the British had great expectations: the Royal Aircraft Factory's SE5. With it came a change of attitude: emulating the French and Germans, the RFC began to concentrate its best fighter pilots in certain squadrons. No. 56 was the first. Its commander, Major R. G. Blomfield, had the sense of style that consorts well with membership of a select community. He valued the Epicurean niceties of life, even in wartime. While the squadron was spending six weeks at London Colney, taking delivery of its aeroplanes and working up to operational standard, he formed an orchestra from the rank and file of his squadron, to play in the officers' mess every evening. Among his pilots was Ball, who had been sent home to instruct towards the end of 1916.

Now very much the seasoned campaigner, Ball expressed forthright views on his craft. The SE5 had a top speed of 120 m.p.h. at 6500 feet and was a strong but light machine, with a 400-round Vickers firing through the propeller and a Lewis on the upper wing, provided with 4 double drums of 97 rounds. Its two and a half hours' endurance was one hour better than the Albatros's, which meant that its pilots had ample time to stay at maximum height waiting to surprise enemy fighters. Not all pilots took to it immediately. The first objection was to the windscreen, which they said obstructed their view and became blurred by scratches and oil. The Lewis gun, despite its Foster mounting, was inordinately difficult to reload on account of wind resistance.

In a letter, Ball denounced it as "a dud" and alleged, exaggeratedly, that its speed was only about half a Nieuport's. This sounds like adolescent petulance rather than a rational assessment. In fact the SE5 was 15 m.p.h. faster than the Nieuport 17. He told his parents he was "making the best of a bad job" by having the Lewis gun taken off to save weight and the windscreen lowered to cut down wind resistance. Some pilots had the windscreen removed. They were all given a free hand about modifications: "But it is a rotten machine."

Familiarisation changed his, and others', opinion. Gunnery practice showed that it was a steady platform from which to shoot accurately. It was nimble in aerobatics. Among battle-hardened young men of eighteen and nineteen, or in their very early twenties, who owed their lives to their above-average skill as pilots as well as to their marksmanship, competitiveness was rife. Someone decided that he would make his landing more interesting and smoother if he touched down on a hangar roof instead of the ground and rolled down it onto the airfield. Soon they were all doing it. When they took off to fly to France on 7th April they went as masters of what could be the most formidable single-seater at the Front. Its imperfections were the erratic Constantinescu-Colley synchronisation gear for the Vickers gun and the temperamental Hispano-Suiza engine.

The Sopwith Pups were still doing respectably against the Albatroses and Halberstadts and the RNAS was giving the RFC valuable help. As an instance, five Pups of its No. 3 Squadron shot down four Halberstadts in a fight on 6th April. The squadron had been formed in November the previous year to meet Trenchard's request for reinforcement. It was commanded by Squadron Commander Redford "Red" Mulock, who, like the majority of his pilots, was a Canadian. Among these was Raymond Collishaw. He had been posted to Naval Three when the wing to which it belonged had been ordered to transfer its nine best fighter pilots; of whom five were Canadians.

After serving with a squadron that had been escorting bombers on average once a week, Collishaw and the others were "rather jolted by the first few patrols" in their new sector. Having been on operations for some months, held their own against enemy fighters and come to regard flak as routine, they were a trifle over-confident as well as unprepared for the intensity of their new duties: which entailed daily patrols and frequent encounters with Richthofen's Jasta 11. Their offensive patrols were usually flown in formations of four or five aircraft at 12,000 to 16,000 feet, where their guns often froze. They also escorted bombers, and FE2bs on photo recce. The bombers were often BE2cs which Collishaw said "could stagger up to 5000 to 8000 feet and cruised at a snail's pace" and he could not understand "the crass stupidity" of those who kept ordering these machines to be built. One escorting flight flew close to the aeroplanes they were protecting and another flew above. Sometimes a third flight gave top cover. It was difficult to persuade the escorting pilots to stay in formation when attacked: they preferred to break and fight. The enemy always attacked the upper flight first. This became known as the "sacrifice flight"; and, if there were a top cover, that was "super sacrifice flight". The newest pilots, being considered expendable, were usually given these positions.

There was yet another Sopwith fighter of which the enemy was wary: the Triplane, or "Tripehound". This had come into service with the RNAS the previous year. On 27th April Collishaw was posted as commander of B Flight to Naval 10, which flew it. Its prime assets were the manoeuvrability and rate of climb imparted by its three wings. It was armed with one, sometimes two, Vickers synchronised with the airscrew. At 10,000 feet, which it could reach in 12 minutes, its speed was 110 m.p.h. Its ceiling was 20,000 feet. Its narrow wings allowed the pilot an excellent view. Richthofen thought it the best Allied fighter during the first six or eight months of 1917.

Collishaw found, to his anger and disgust, that there were pilots on the squadron who "could not, to put it charitably, be depended on. Some of these were merely inept, but others simply did not want to fight." He quickly found himself abandoned by most of his flight in a fight. The other flight commanders suffered the same desertion. The reason for this cowardice was that when other squadrons had been asked to contribute to the formation of the new No. 10, their commanders had sent those they valued least. Collishaw persuaded his squadron commander to rid himself of the faint-hearted. Most of the replacements, and all the pilots on Collishaw's flight, were Canadians.

It has been suggested that Collishaw and his pilots decided to paint their Triplanes black all over "because they would look murderous". The truth is that all the squadron's aeroplanes had a khaki-green upper wing surface and fuselage, with pale blue on the wings' undersides. To distinguish the flights when airborne, the CO decided that A Flight's cowling, top and sides of the fuselage, and wheel discs would be painted red; B Flight's black; and C Flight's blue. What Collishaw did suggest was that B Flight should name its aircraft. He called his *Black Maria*. Flight Sub-Lieutenant Ellis Reid chose *Black Roger*. John Sharman fancied *Black Death*. Gerry Nash named his *Black Sheep*, and Alex Alexander's was *Black Prince*.

This touch of flamboyance was justified by B Flight's performance, as we shall see later. Collishaw scored his first "flamer" when he shot down an Albatros. It "was not at all a pretty sight but there was the comforting thought that it was far better to have happened to him than to me". But that was on 10th May and we have not done with Bloody April yet.

It was not until 22nd April that 56 Squadron was ready for action. Then, on its first patrol, it shot down four Albatroses, of which one fell to Ball. The pace was at once frenetic and the stress unremitting. On 29th April Ball was already writing to his parents: "I am so very fagged. April 26 evening I attacked four lots of Huns with fire. Brought two down and had to get back without ammunition when dark. Had four fights and got one Hun. In the end all my controls were shot away. But I got back. Simply must close for I am so fagged." This was hardly the sort of letter that a mother or father would relish receiving. But Ball was not gregarious and his parental tie was uncommonly close. Telling them the truth about the dangers of his life must have been his only outlet to relieve nervous tension.

Mannock was on his first tour at the Front, where he had joined 40 Squadron on 7th April, to fly Nieuports. The immediate impression he made on his new comrades was disastrous. Shyness, added to a social clumsiness that fitted him for the barrack room and sergeants' mess rather than for the company of officers, and, on top of those handicaps, his eagerness to be informed in detail about his new environment, led him into a performance that presented only the obnoxious facet of his personality. The CO, Major Leonard Tilney, took him to the mess and introduced him to other members of the squadron; most of whom had just landed from a patrol on which Lieutenant Pell, a most popular figure, had been killed. The atmosphere was muted. Mannock was impervious to the melancholia and bumptiously asked everyone how many Huns he had fanned down: a highly offensive inquisition in any circumstances. He then plunged into a dissertation on air fighting and expressed his views on the war in general. To compound his grossness, he had seated himself in the chair usually occupied by Pell. Also, he had been fortunate enough, when training, to acquire more flying hours on tractor types than any of his new companions – who had until recently been flying pushers – had accumulated in their considerably longer service. This did nothing to endear him.

A fellow pilot, recalling Mannock's seemingly conceited, bombastic self-introduction, recalled: "Apart from that, he was different. His manner, speech and familiarity were not liked. He seemed too cocky for his experience, which was nil. New men usually took their time and listened to the more experienced hands. He was the complete opposite and offered ideas about everything: even the rôle of scout pilots and what was wrong with our aeroplanes. He seemed a boorish know-all."

Hubris suffered its classical chastening. After practice firing at a ground target, Mannock expected no difficulty in shooting down the enemy. Six days after his arrival, he flew his first operation, as part of an escort to some FEs. Nervousness caused him to keep mishandling his throttle and losing his place in the formation. He did no better on his next few sorties. His reactions were so slow that he was always the last to go into a fight and was soon suspected of cowardice. Treated coldly, snubbed, he endeavoured to ingratiate himself by chipping into conversations; and, an unforgivable breach of mess custom, expatiating on politics: his left-wing bigotry, of course, exacerbated the transgression against good form.

He was also ridiculed behind his back – the others ignored him – for his persistent gunnery practice. As Ball had always done and McCudden and others of the greatest fighter pilots would emulate, he loaded his ammunition drums himself and spent hours making repeated dives at a ground target at different speeds and from different angles. To his detriment, however, he could not refrain from trumpeting about it and announcing that if he could open fire from twenty yards he would not miss. This provoked further dislike.

On 1st May his flight escorted four $1\frac{1}{2}$-strutters bombing Douai aerodrome, where Richthofen's Jasta was based. When he tried to clear his Lewis gun, en route, it jammed. Fearing the squadron's scorn if he turned back, he flew on armed only with his revolver. He heard machine-gun fire astern, looked round, and saw a yellow-and-green Albatros diving at him. He turned tightly to face it. Talking about it afterwards he said he heard a strange noise above his engine; and realised that he was screaming with anger, which helped his nerves. (If ever a man needed psychiatric help, he seems to have been a prime candidate.) The German, seeing Mannock hurtle at him, broke and turned on another Nieuport. But this brave refusal to be daunted by a hated, despised German did not improve Mannock's standing with his fellows.

On 7th May he attacked his first balloon. Six Nieuports went five miles behind the enemy lines at fifteen feet, to put into practice a new form of balloon attack perfected by Tilney. He had thought out this low-level approach to outwit the enemy: who were no longer winching down their balloons when attacked, but bringing them to ground much more quickly by passing the cable under a pulley and attaching it to a lorry; which then simply drove off fast. While they scudded through successive belts of ground fire, Lothar von Richthofen, Manfred's young brother, led five Albatroses up over the balloon line to wait for them. Captain Nixon, who led the Nieuports, indicated a target to each pilot. Mannock's was at the far end of the line. He fired long bursts of tracer into it. It crumpled and began to fall. Mannock turned for home. Meanwhile Nixon had spotted the Albatroses and peeled off to try to protect his men. This drew the enemy off. They milled around Nixon and it was not difficult, with four others to distract the victim, for Richthofen Minor to shoot him in the back and bring him down.

Captain Nixon's gallant death was little noticed outside his squadron and his family. But 7th May 1917 was memorable to all the air forces at the Western Front and the whole British nation for another sad event.

On 3rd May Albert Ball had written home. "It is quite impossible, but I am doing all I can. My total up to last night was thirty-eight. I got two last night. Oh! It was a topping fight. A few days ago all my

controls were shot away on my SE5. But I got the Hun that did it. I[
is all troubles. I am feeling very old just now."

Two days later: "Dearest Dad, Have just come off patrol and made
my total forty-two. I attacked two Albatros scouts and crashed them
killing the pilots. In the end I was brought down, but am quite O.K
Oh! it was a good fight and the Huns were fine sports. One tried to ram
me after he was hit, and only missed by inches. Am indeed looked after
by God, but oh, I do get tired of always living to kill, and am really
beginning to feel like a murderer. Shall be so pleased when I am
finished."

On that day, talking to Atkins, another pilot, Ball said: "Trenchard
says I can go home when I have got fifty. But I shall never go home."

Two days after that, the 7th May, in the evening, eleven SE5s of 56
Squadron took off on the day's last patrol. Behind enemy lines at 18,000
feet, they saw six Albatroses 3000 feet below and dived to attack. A mile
or so away, two Jastas were approaching. When they saw the British
begin their attack, they followed them down and opened fire on them
from the rear. The dogfight, in which forty-one aircraft were involved
sprawled all over the sky and down to 600 feet before the last contestant
departed for his base. Six SE5s went down, Ball among them. He was
last seen on the tail of an Albatros. To this day it is not known for
certain whether the six Albatroses that drew 56's attention were the bait
in a trap. Nor has it been established that it was Lothar von Richthofen
who killed Ball. There is a legend that he did. But the only claim he
made on that date was for a Sopwith Triplane. Ball's confirmed victories
totalled forty-three and no one can say how many more he shot down
in that last fight.

He was awarded a posthumous Victoria Cross "For most conspicuous
and consistent bravery from 25th April to 6th May 1917, during which
Captain Ball took part in 26 combats and destroyed 11 hostile aeroplanes
drove down two out of control, and forced several others to land".

When Ball had left 60 Squadron to return to England in late 1916
already, at the age of twenty, a captain and decorated with the DSO and
MC, he had been thought irreplaceable and his record impossible to
equal leave alone surpass. And he was irreplaceable until a twenty-two
year-old Canadian, Lieutenant William Avery "Billy" Bishop, joined
the squadron on the very day of Ball's death. At the war's end Bishop
had seventy-two victories; only one less than Mannock, the RFC's most
prolific destroyer.

Bishop's way thither had been meandering, and, in a manner, adven-
titious. In 1911, aged seventeen, he entered the Canadian Royal Military
College; and barely managed to stay there for the next three years: his

frequent mischievous escapades courted expulsion. But, if he was not by temperament a conventional military or naval type, it was in his stars to become one of the greatest fighter pilots the world has seen; so he was lucky to be born at the right time to give full vent to his natural aptitude and fulfil his destiny. A spice of devilment and flouting of regulations may not engender the right chemistry to take a young man up to the rank of general or admiral; but they have set many a regardless junior officer on his way to an air marshal's pennant.* Billy Bishop emphatically was one of the best pilots of an earlier, or any, war. And if he had given up breaking rules for the hell of it when he passed out from the Academy, his independent attitude remained undiminished when he eventually put up his pilot's wings and reported for duty on the Western Front.

First commissioned in September 1914 in the 9th Missisauga Horse, Bishop transferred to the 7th Canadian Mounted Rifles and left Canada in June 1915. Like many others before him he found the mud in a cavalry camp unpleasant. And, as many others had been, he was attracted to the notion of flying only when an aeroplane landed nearby. Told that the quickest way to start operational flying was as an observer, that was what he volunteered to be. After a course at the observers' school he put up his half-wing and went to France in January 1916 to an RE7 artillery co-operation squadron. He wanted to be fighting, not observing, but admitted that it was "no child's play to circle above a German battery with your machine being tossed about in the air, tortured by exploding shells and black shrapnel puffballs coming nearer and nearer to you". After four months he injured his knee in a crash and was sent to hospital in England, where he was grounded until November, when he was sent to flying school.

In December he joined No. 37 (Home Defence) Squadron, with whom he did a lot of night flying, "a fearsome thing but very interesting". Bored, and thirsting for a fight, he volunteered for France and arrived there on 7th March 1917 to fly Nieuport 17s with 60 Squadron.

On his first patrol, flying at the rear of a six-machine formation, and given, as usual with newcomers, an old aeroplane that was slower than the others, he had trouble keeping pace and was in constant fear of being separated. This in fact happened. A shell burst so close to him that he lost 300 feet and was left half a mile behind. He managed to rejoin, but, in common with other pilots new to battle, he found, after a few manoeuvres, that he had lost sight of his companions.

* Jim "Ginger" Lacey, the RAF's top-scoring pilot in the Battle of Britain, who became a squadron commander, has said: "It's always the best pilots who break the rules and get into the most trouble."

Presently he saw them below, diving, and hurried after them. There ensued an episode of the sort that was peculiar to the first war in the air and lent welcome light relief to the fear and tension of the daily business of killing and being killed. The object of the Nieuports' attention was a large white German aircraft that daily did artillery spotting in the same patch of sky. This familar object was known as "the Flying Pig" because it was decrepit, slow and crewed by an incompetent pilot and observer. It was a point of honour in the squadron that the bumbling antique contraption should not be shot down. It was considered fair sport, however, to frighten it. Bishop said: "Whenever our squadron approached, the Pig would begin a series of clumsy turns and ludicrous manoeuvres and open a frightened fire from ridiculously long ranges." The observer was a very bad shot and never succeeded in hitting any of the British machines. So attacking this particular German was always regarded more as a joke than as a serious part of warfare. "The idea was only to frighten the Pig, but our patrol leader had made such a determined dash at him that he never appeared again. For months the patrol leader was chided for playing such a nasty trick on a harmless old man."

But these lighter episodes were few and on 25th March Bishop figured in one of a very different sort that was much more typical of life at the Front. Patrolling with three others he saw three Albatroses which the patrol leader pretended to ignore. The Albatroses accordingly fell in astern and chased the Nieuports. When the distance between the two had narrowed to 500 yards, the Nieuports turned sharply about to engage the enemy. Bishop saw that his shots were hitting one of these, which promptly half-rolled, then began to spin. Suspecting that it was departing the scene rather than going back damaged, he followed it, shooting every time it pulled out of its spin; until it crashed. In the long dive Bishop's engine had oiled up and stopped. He thought he was behind the German lines, but had no alternative to a forced landing. Luckily he was 150 yards behind the British trenches.

He had made his first kill, and two days later he led a patrol for the first time. He acquitted himself less than well. Seeing a single hostile below, he signalled to his pilots and dived on it. A second hostile appeared from cloud. When tracer from a third started streaking past him from astern he realised he had been duped. Meanwhile more enemy aircraft had appeared out of the clouds and attacked the rest of his formation. He looped and rolled off the top to avoid his three adversaries, then saw another fight in progress, which he joined: two Nieuports against four Albatroses. This went on for fifteen minutes without either side scoring a kill. He broke off to help another Nieuport that was on its own against two Albatroses. This combat brought them down to

2500 feet before the enemy broke off. Two Nieuports failed to return from this patrol.

Of one of the pilots who was killed, Bishop said: "He was only eighteen and had been in France three weeks. The RFC is filled with boys of that age with spirits of daring beyond all compare and courage so self-effacing as to be a continual inspiration to their older brothers in the Service." He was hit in the back by an explosive bullet that exploded in his stomach. He continued fighting for ten minutes, then crash-landed and fainted; and died in hospital.

On 7th April Bishop was ordered to attack a balloon but forbidden to descend below 1000 feet. It was being winched down, so he dived, began firing at 500 feet and continued doing so until he was only 50 feet above the ground. Throughout, he was under fire from "flaming onions", which he described as "terrifying balls of fire shot from some kind of rocket gun". They weren't. They were shells from 37-millimetre cannon which, seen from the air, seemed to come spurting up in clusters of incandescent globules. His engine cut but caught again when he was down to fifteen feet, and he flew home through machinegun fire. He received congratulations from the Wing and Brigade Commanders and a telegram from Trenchard.

The rest of the month was a succession of patrols and ground strafing. On the 25th he was promoted to captain and by the end of Bloody April had shot down at least twelve enemy aircraft.

At the beginning of that catastrophic month the RFC had 754 aircraft, of which 385 were single-seater fighters, at the Front. The Germans opposed them with 114 single-seater fighters.

Thirty days later 151 British aircraft had been brought down, compared with 119 German.

The RFC lost 316 aircrew, dead and missing. The Germans lost 119.

The expectation of life for the British had averaged twenty-three days.

CHAPTER 15

———————◆———————

1917. The Darkening Days

"The days darken round me, and the years,/Among new men, strange faces, other minds" Tennyson wrote.

Strange faces had been seen on the squadrons from the moment the first shot was fired. Even before that, the constant replacement of those killed in accidents made new men a familiar feature of life in any air force. Death in action was a different matter from death caused by engine failure, a stall, lost sense of balance and orientation in cloud, or an unsuspected hill looming suddenly through the mist. Every pilot believed that he could avert an accident. But every pilot who saw a comrade killed by bullets or shellfire brooded on what he himself could have done to avoid the same end in exactly the same circumstance and had to conclude that the answer was: "nothing". Observers could only hope or pray that their pilot was born under a lucky star. Other minds devised new tactics forced on airmen by the changing times.

By now, with the war almost three years long, the days had indeed darkened. The raw, often squalid existence, haunted by fear, of men teetering on the brink of mental and physical exhaustion that has so often been portrayed as typical of squadron life at the Western Front had become a reality on both sides of the battle line. The average officers' or sergeants' mess was cramped and its furnishing shabby. Despite the close comradeship, men whose nerves were stretched almost to breaking point were constantly irritable and intolerant of some of their companions' mannerisms and characters. Many tried to avoid forming close friendships and became remote and introspective because they knew the grief that lay in wait when an intimate friend was killed. Drink became the solace and support of some. There were a few who resorted to the whisky or brandy bottle before every sortie.

This is not to say that a visitor to any mess found it a dingy hovel reeking of alcohol fumes, with scruffy, surly, dispirited pilots and observers sprawled in an intoxicated stupor on sagging armchairs with broken springs and stuffing protruding through rents made in the course of robust nightly revelry. Far from it. Stress and anxiety were mostly betrayed by an air of ineffable weariness, a tendency to fall into a doze at any odd moment, a morose taciturnity; and a variety of tics: head-jerking, shoulder-shrugging, blinking, trembling hands, foot-tapping. These were the signs for which squadron and flight commanders and medical officers looked. Often it was a squadron or flight commander himself who showed symptoms of excessive wear and tear on mind and body.

One cause of increased danger and taut nerves was the low-level ground attack that the April offensive had brought into prominence. The days when this had not been so risky a task as had been expected were gone. The men in the trenches had learned how to defend themselves against strafing from the air. Heavy machineguns and light anti-aircraft guns protected by sandbagged emplacements abounded. Even volleys of rifle fire were shooting down strafing aeroplanes.

Danger did not mesmerise men into a state of permanent dejection. The French tended to seek their release in a local town and short leaves in Paris. The Germans, with their customary arrogance and thick skins, imposed themselves on the restaurants and brothels of Lille, Douai and Metz or kept to themselves in their own messes. In the RFC the common antidote to an oppressive sense of imminent death or disabling injury was boisterous entertainment of other squadrons and occasional excursions to St Omer, Amiens, Béthune or wherever they could do some shopping and enjoy a better meal than their own cooks could provide. Bridge and poker were popular, the latter particularly with Canadians and Americans. Gambling was seldom heavy: Service rates of pay took care of that.

Collishaw, who had a great reputation for enlivening a party, described how the members of the squadrons on which he served worked hard and played hard. One of the main objects on guest nights was to ensure that the hosts regretted having invited their visitors. Messes were usually large wooden huts with no ceiling. A popular diversion was to jump up to grab the rafters and make concerted attempts to jerk them down and collapse the roof. These often succeeded.

On one occasion, he says, when his squadron entertained No. 32, the guests were given the Visitors' Cocktail, a tall glass containing everything in the bar. At the end of the revels, "we stacked them in a lorry like firewood and sent them home". The next day, Brigadier General Longcroft, the Brigade Commander, turned up, did not specify his purpose, wandered about the hangars, peering speculatively about, then asked Collishaw's squadron commander if his pilots had cracked up any aeroplanes that morning. Assured that there had been no accidents, the Brigadier asked if the dawn patrol had got off all right. "Of course," said the CO, "why not?" Longcroft explained that he had been to 32's aerodrome earlier and found the pilots in a sorry state. They had tried to send up a dawn patrol but two had crashed and the take-off had been abandoned. Few of Collishaw's squadron had been fit to fly, "but the CO had scraped together a few hardened drinkers" and the dawn patrol had duly been carried out.

He also said that when his squadron was at Luxeuil with the Escadrille Lafayette there were usually three poker games going on: one with a five-franc limit, another with a twenty-franc limit and the third with no limit. This last was the one that attracted most Canadians and Americans, each of whom would bring a roll of lavatory paper to the table on which to sign IOUs.

Another diversion at Luxeuil when the weather turned freezing cold and the airfield was a morass, was provided by flooding a hangar to form an ice floor. On this, the British, Canadians and Americans played hockey. They had their own rules. "If a player was within ten feet of the puck he was fair game and could be tackled." Such sport as this held no more appeal for the Aviation Militaire or Luftstreitkräfte than did playfully wrecking other squadrons' messes as an expression of the greatest of good will and mutual esteem.

Carousal was not a nightly event. Collishaw said that everyone was too tired. Flying six or seven hours a day was "a dreadfully fatiguing business" and even though they had all the energy and resilience of youth, it left them "too drained physically and emotionally to carry on late into the night", unless bad weather prevented flying. The only break for the pilots on an RNAS squadron was the Navy's traditional "make and mend" weekly half holiday. "The sensible thing, of course, would have been to spend the brief period in relaxation and roll into bed at an early hour. That was a little too much to expect of a group of young men such as we were." What they did was rush through lunch and then make for the nearest large town for a "bangup" dinner and whatever other entertainment, "respectable and otherwise", was on offer. They would return to base "befuddled but happy", snatch a couple of hours sleep, then be roused for dawn patrol. "The onrush of cold fresh air into the open cockpits had a remarkably sobering effect." Later generations of pilots would gratefully clamber aboard with a hangover and switch on the oxygen to obtain the same therapeutic effect.

If any Allied pilot relished killing as much as Richthofen, it was Billy Bishop. But whereas Richthofen enjoyed killing for its own sake, Bishop enjoyed killing only Germans. He hated them because they were the aggressors who had deliberately started the war. He also despised them.

Collishaw declared himself not to be one of those few who hold the view that war is an ennobling and welcome experience. Only a psychopath or person without pity or principles who sees personal profit in war could hold such an opinion, he maintained. Bishop felt the same. Richthofen plainly did not and represented everything that these two sane young Canadians abhorred.

Because he was killing the enemy, Bishop had enjoyed himself in April and he added to his enjoyment in the following few months. On 25th April he had been made a captain and given command of a flight; so was able to go off on his own when he felt like it, to emulate Ball and the other great British and French loners and try to exceed their scores. His excuse was that he preferred not to have others' lives in his hands: a poor one for a professional officer at any time and for a flight commander hardly a good qualification. Typically, when on contact patrol and ground strafing he saw infantry in attack being slaughtered by machinegun and artillery fire, it exacerbated his hatred of the Germans. One day when he saw his own troops being cut to ribbons and forced to stop and drop prone, he spotted a group of the enemy in a corner of a trench manning two Spandaus. He dived vertically at them, shooting. One of the guns returned his fire. From a height of thirty feet he "could make out every detail of the Huns' frightened faces. With hate in my heart I fired every bullet I could at them." A few minutes later he saw the British troops advancing again.

"Soon after this new Hun-hatred had become a part of my soul," he attacked a two-seater which presently turned tail and landed. To see the German alight under perfect control "filled me with a towering rage. I vowed an eternal vendetta against all the Hun two-seaters in the world." He dived to within a few feet of the ground, firing a stream of bullets into the machine. "I had the satisfaction of knowing that the pilot and observer must have been hit or nearly scared to death." He waited a while, but nobody emerged from the riddled aeroplane. Half an hour later he saw three Albatroses in combat with a Nieuport and shot one of them down with a short burst. The other British pilot sent another down. The third Albatros departed.

Another day he saw three two-seaters artillery spotting and attacked them. Each fired a few rounds at him with both its guns, then they all fled. He followed and soon saw five scarlet Albatroses overhead. These belonged to Richthofen's Jasta, No. 11. He climbed above them, and while the two-seaters drew off he made a series of dives on the fighters. Their pilots had expected him to dive through them, but instead he zoomed when still over their heads, to make a series of diving attacks. This confused the rigidly drilled enemy, who scattered. He then set off to find the two-seaters, incensed because although they outnumbered and heavily outgunned him, they had refused combat. He shot one down, killed the observer in another and the third got away altogether.

In May Bishop was awarded the MC and in June the DSO.

He planned a single-handed attack on Jasta 5's aerodrome at Estourmel, and carried it out on 2nd June. He was called at 3 a.m. and took off shortly before five o'clock to hedge-hop his way to the target. Having climbed to 200 feet as he approached, he saw six Albatros DIIIs and one two-seater outside the hangars, so dived and strafed them. Two Albatroses were taking off and he shot them both down. Two more followed them. He shot one down and attacked the other but had emptied his ammunition pan. Rather than hang about while he reloaded, knowing that other enemy fighters might appear at any moment, he made off. Thirty-seven minutes after having set off, he landed. Trenchard extolled this feat as "the greatest show of the war" and put him up for a VC; which was gazetted in August.

Bishop's action had confirmed Trenchard's theory. Trenchard deplored the adulation of fighter pilots for the number of enemy aircraft they brought down. He did not make this public, but the view he expressed in private was that bombing made a far more valuable contribution to winning the war. And, he said, the right time to destroy enemy aeroplanes was before they could get off the ground. Then, the greatest damage could most economically be inflicted. The attack that Bishop thought out for himself could not have met with greater approval from Trenchard if he had personally briefed Bishop for it: the destruction of aircraft on the ground; in this instance, by a fighter pilot using only a machinegun.

Collishaw, that altogether admirable man, has been given far too little recognition by air historians. They are mostly concerned with British, French and German fighter aces and tend to ignore those from the Commonwealth; except Bishop, who couldn't be overlooked with only one victory less than the top-scoring Mannock.

Collishaw had a pithy view to express, which coincided with Trenchard's. Newspaper reporters who visited aerodromes ignored all but the fighter pilots, he pointed out, because there seemed to be an inherent glamour in their rôle. The work of artillery observation pilots, for instance, was highly technical and made good reading only when something went wrong. The public derived its information from *The London Gazette*, in which awards were published. A decoration for "Fearlessly attacking a large number of enemy aircraft" made better reading than "carried out X number of photographic reconnaissances".

Air fighting had not developed for its own sake but to prevent the enemy from operating over one's lines and to protect bomber and reconnaissance aircraft.

Lloyd George was considerably to blame. In the summer of 1917, letting off verbal wind in the House of Commons on the subject of fighter pilots, he brayed: "They are the knighthood of this war, without fear, without reproach; and they recall the legendary days of chivalry, not merely by the daring of their exploits but by the nobility of their spirit." Referring to these specious platitudes, Collishaw condemns them as nonsense: sport was confined to jousting, but when knights went into battle they fought to kill. Where was the chivalry?

With a character like Richthofen roving the sky like a mangy but dangerously rabid jackal, ready to snap up any Allied aeroplane of inferior performance and armament to his own, and often protected by his admiring henchmen, who held back to let him add to his shabby tally and keep at bay any of the victim's friends who might try to help him, sloppy references to chivalry are totally ludicrous. The essence of combat, in the air, on land or at sea, is exemplified in Bishop's shooting up the two Germans when they were defenceless on the ground after ducking out of a fight. It wasn't sporting but it made excellent sense. Had they voluntarily landed on the Allied side of the lines, he would have left them alone. As it was, why let them live to fly another day and kill an ally?

At the same time as Bishop and Collishaw were terrorising the enemy, Mannock was still striving to establish himself in the regard of his fellows. On 9th May he was alone behind the German lines at 16,000 feet after his leader had turned back with engine trouble. Three Albatroses appeared and he went for them but his gun jammed. He spun out but they followed him down. In a succession of spins punctuated by brief moments of level flight, when the enemy would resume shooting, which took him down to 6000 feet on his own side of the lines, the Albatroses withdrew. He at once recrossed the lines but found no trade, so went home. Major Tilney, seeing how tired and dejected he was, gave him a respite. He was due to patrol again in the afternoon, but Tilney sent him as a passenger in a 16 Squadron RE8 to fetch a spare Nieuport from St Omer. On the way he underwent a change of attitude. Thinking over the intense fear he felt when over Hunland or in combat, he realised that he must overcome it and apply himself entirely to "learning the game". Thenceforth he was noticeably more relaxed and cheerful and the squadron treated him with increasing friendliness.

But he was still making no reputation as a fighter pilot. Sortie after sortie produced no success, although he unhesitatingly attacked every enemy aircraft that came within range. Major Tilney was intelligent enough to divine that Mannock needed encouragement and sympathy and a display of confidence in him. He therefore often let him lead a patrol. This increased his confidence but did not improve his luck. On 25th May he led his flight into a fight with two artillery spotters. One was too high to catch but Mannock fired and frightened it off, then turned at the other and shot thirty holes in it around the area of the pilot's seat. The aircraft nosed down slightly and flew on. Mannock was convinced that he had killed the pilot and that the aeroplane was merely continuing as it had been trimmed to do. Nobody had seen what happened, so he did not claim a kill for fear of disbelief and ridicule.

He was shooting and missing so often that he began to worry about what was wrong with his aim. As he had an almost blind left eye, one would have supposed the answer to that was obvious. On 7th June he shot down an Albatros at last; his first confirmed kill. Five days afterwards he was coming in to land when he felt intense pain in his right eye. He fainted under the medical officer's ministrations. After the eye had been anaesthetised in hospital the surgeon removed a large piece of grit and a sliver of metal from it. On 17th June he was sent home on two weeks' leave; where he found that his mother had become an alcoholic. He was glad to return to France.

On 12th July he shot down a DFW two-seater on the Allied side of the lines and went to see the wreckage, to check the accuracy of his shooting. He was nauseated by the sight of a small black dog in the observer's cockpit, pulped by his bullets, and the two dead men; the pilot a pulverised mess of broken bones and torn flesh. But he found three bullet holes in the side of his head. The following day he sent down two more two-seaters. Until then some of the squadron had still doubted his courage.

A newcomer to the squadron at about this time affords us a welcome different view of Mannock from the old stagers'. On the afternoon when McLanachan arrived he was astonished to find a tennis court, where a game of doubles was being watched by officers in deck chairs. Three of the players were actually wearing white flannels. His reception was casual but friendly. Asking where the CO was, he was told: "That's the CO, over there, but you'd better not disturb him until the set is finished."

Noticing him, Major Tilney, "a young, rather florid-faced youth", called: "You the new pilot? See you when we've won this set."

The set over, Tilney came to chat and asked McLanachan if he had ever flown a Nieuport. He had. To his surprise Tilney called to the others: "Come on, here's a fellow from Smith Barry's squadron who has flown a Nieuport. Let's see what he can do." Smith Barry's work as an instructor on No. 1 Reserve Squadron, which he commanded, was already well known.

So up McLanachan went and at 1000 feet the engine cut. There was not enough height to dive and restart the engine, so he decided to perform one of No. 1 RS's favourite stunts and do a spinning nose dive: which, everywhere but in Smith Barry's unit, was regarded as a death trap. He landed safely and neatly in the middle of the airfield and waited for the mechanics to start the engine so that he could resume his display. Instead, one of them told him that the Major would like to speak to him. "Much to my amazement, on approaching the CO he turned away from me and walked off." Hurrying after him, McLanachan began to apologise, but Tilney waved him away.

A tall weather-beaten pilot was laughing. This was Mannock, who explained that all the onlookers had thought he was going to crash and kill himself. "We don't like watching fellows kill themselves, and Tilney looked away when he thought you were finished."

McLanachan instantly feared that Tilney would send him back for further training, but Mannock reassured him. "If you can handle a machine like that we want you in this squadron."

Ever afterwards, said McLanachan – whom Mannock dubbed "McScotch" – he felt grateful to him for these friendly and encouraging words. They saved his self-respect and showed him that Mannock had "a somewhat Puckish sense of humour". Everything about Mannock, according to McLanachan, demonstrated his vitality: a strong, manly appearance and directness of speech. They became friends immediately.

It was not many days before McLanachan, on patrol with two others, saw a Nieuport catch fire. It was flown by a nineteen-year-old of whom everyone was very fond. The spectacle was particularly revolting because the Germans had recently begun using incendiary bullets, which were forbidden by the Geneva Convention except against balloons. McLanachan was so angry that before his next sortie he told a mechanic to fill his drums with tracer, armour-piercing and incendiary rounds. The mechanic replied that if he were forced down on the other side with incendiaries in his drums he would be shot and he, the mechanic, court martialled. McLanachan then went off to fill the drums himself. Mannock found him at it, and, looking distressed, reasoned with him.

"They've never fired anything at me *but* incendiary. Could you coolly fire that muck into a fellow creature; or worse still, into his petrol tank, knowing what it must mean?" And Mannock talked him out of it.

Captain Maurice Baring spent his time accompanying Trenchard on visits to the squadrons, by air because lack of time demanded it, and to Wing and Brigade HQs. He made the interesting comment in his diary that Harvey-Kelly said the Germans he met in the air now were like floating meat. A month later: "We got the news that Ball is missing. This cast a gloom through the whole Flying Corps. He was not only perhaps the most inspired pilot we have ever had, but the most modest and engaging character." Between these events, on 29th April: "We went to Vert Galant to see Harvey-Kelly, who commanded No. 19 Squadron. When we got there we were told he had gone up by himself and one other pilot for a short patrol." Then he wasn't by himself; and he went *with* another pilot. "By lunch time he had not come back. He was due and overdue. When we went away the General said: 'Tell Harvey-Kelly I was very sorry to miss him,' but I knew quite well from the sound of his voice he did not expect this message would ever be delivered. Nor did I.

"Harvey-Kelly never came back. He was the gayest of all gay pilots." (Would one dare write that now, intending high praise?)

But there were compensations. The 4th June, their Founder's Day, was celebrated by 300 Old Etonians at a dinner in the Lord Roberts Memorial Hall. (He doesn't say where; St Omer? Ypres?) "I knew about five by sight. All my contemporaries were Lieutenant-Generals. They sang, accompanied by the Coldstream Band, and after dinner everything in the room was broken: all the plates, all the glass, all the tables, the chandeliers, the windows, the doors, the people . . ."

Collishaw was finding little time for rowdy evenings. He had an experience that almost matched one of Louis Strange's two years earlier. In a dogfight at 16,000 feet he was turning inside an enemy machine and keeping an eye on another that was trying to get on his tail. A third Albatros loomed dead ahead. There was no time to consider stress limits. He shoved the stick right forward and just scraped beneath the aircraft in front of him. The violence of the manoeuvre broke his safety strap. He found himself outside the cockpit. He grabbed at the two centre struts. The engine was at full power and the controls were left to their own devices. The Tripe dived steeply with Collishaw's legs and lower trunk trailing in the slipstream. Suddenly its nose reared up, then it stalled into a falling-leaf spin. It took all his strength to hang on. His hands were slipping when the aircraft pulled up sharply and his body slammed onto the cockpit coaming. He worked one leg into the cockpit and hooked a foot round the stick to pull it back and settle the Tripe in level flight. When he fell into his seat he was down to 6000 feet.

In the months of June and July, his flight shot down eighty-seven enemy aircraft for the loss of one pilot.

On 4th June Henderson wrote a summary of the work and worth of the Royal Flying Corps, for the consideration of the Chief of the Imperial General Staff: excellent in content, rambling in style. He does not refer specifically to the events of April, but the allusion to them, couched with modesty and dignity, is inescapable.

"When the Royal Flying Corps crossed to France in 1914, it held certain advantages and suffered from certain disadvantages connected with the aerial forces of other nations. In the first place, it was well organised, the personnel had been carefully selected, and both officers and men had been trained in their particular duties solely with a view to efficiency in war. The squadrons had come fresh from a camp of instruction at which all the possibilities of the Air Service had been practised, and had been freely discussed by the officers and men. Although there was neither tradition nor war experience to guide them, certain theories had been evolved that in the main proved correct, and above all, the Corps had become imbued with a spirit of duty and devotion which was quite remarkable in so young a service. The disadvantages under which it suffered were: firstly, the small numbers; secondly, the variety, both in design and efficiency, of its aeroplanes; and thirdly, its limited supply of engines, which in addition were almost entirely of foreign manufacture.

"For the first year of the war these advantages proved so great that, in spite of the early deficiencies, the Flying Corps was able to carry out its duties so successfully that its value as one of the arms of the Service became firmly established. In its action two principles were never departed from, one was that information for the Army must be obtained at any cost, the other that any enemy met in the air, whether in aeroplanes or in airships, must be immediately attacked. The value of the information obtained in the early stages of the war is, of course, known only to few, but when, after the war, the story of these first months is compiled, one of the real surprises in store for the public will be the realisation of the influence of the work of the RFC on the course of the campaign.

"The early disadvantages under which the RFC laboured were in the main due to the conservatism both in civil and military circles of this country, in respect of possibilities of the new service. Money had been provided not very liberally, and was extracted from the Treasury only by the most persistent efforts of those upon whom the responsibility of building up the flying Service had been thrown. The total lack of comprehension of the requirements of war which up to 1914 had characterised practically the whole of the civil population of these islands, and which in an intensified degree was accepted by the politicians to whom the guidance of the country had been entrusted, led to continual misunderstandings as to the necessity for the complete organisation and the complete equipment of the Flying Corps. The general public could see no reason why anything was required beyond a number of good flyers and a number of good aeroplanes, the type of flier being the man who could play tricks over the aerodrome, and the type of aeroplane being that which each individual pilot might happen to prefer; and not least among the obstacles to efficiency was the ignorant criticism of those who exploited the ignorance of the public in order to advance either their own reputations as brilliant thinkers, or the fortunes of some commercial enterprise in which they were interested.

"Any person who has followed the course of the war can now see for himself that the advantages with which the Flying Corps started have been increased, and that its disadvantages have progressively diminished. After nearly three years of continual fighting, with heavy losses suffered under strain which is unknown to other combatants, the spirit of the Corps is as undaunted as ever, and its offensive has never been relaxed; its numbers have increased to an extent which, although never divulged, is known to be very great; its equipment, both in quantity and quality, has made still greater advances; the scope of its work has been extended to such an extent that the Air Service has become an indispensable factor in successful operations. The very bitterness of the outcry which arises whenever there is a suspicion that our superiority in the air is menaced is in itself an indication of the importance which is attached to its efficiency. Even those who have little knowledge of the operations of our Armies in the Field can measure in some way the growth of the Flying Corps by the enormous development of the aeroplane engine industry at home.

"There are those still who think that aeroplanes and pilots are all that are required to make an efficient Flying Corps, but those who have seen our flying men at work, know very well that it is not by numbers alone, or even by mechanical superiority, that success is to be won, but it is the man who tells in the end; and the fighting men of the Flying Corps have to be trained in other things besides flying. Perhaps the most remarkable thing about the Royal Flying Corps is that it has jumped from tens to thousands, the thousands have inherited the same spirit, the same devotion and discipline, as imbued the little band that crossed the Channel on the outbreak of war."

Among the many matters demanding Henderson's attention was one that was a conspicuously far-sighted appreciation of the possible use of air power and a tribute to the progress aviation had made in the past three years. His paper on this, dated 26th June 1917, is as sensible as the one above; and, notable for its improved lucidity and concinnity. It is entirely relevant here as a rider to the earlier paper and an appreciation of how closely the air arm was already integrated with the General Staff's planning for the development of tactics and strategy.

"The proposal to carry considerable bodies of fighting troops in aeroplanes and land them behind the Enemy's lines, is not a new one, and has at various times been considered in all its aspects. The reasons why such a policy cannot be adopted at present are roughly as follows:—

"(1) The design of an aeroplane to carry a useful cargo of men presents very great difficulties, principally because, with the system of construction which has hitherto been used, relative weight increases with the size of the aeroplane when that exceeds a certain definite scale. No aeroplane which could be called a satisfactory flying machine has yet been produced to be capable of carrying thirty armed men.

"(2) The construction of a large number of such machines could not possibly be undertaken in this country, unless the whole of the aeronautical resources were devoted to the task. That would mean that the construction of fighting, reconnaissance, and artillery aeroplanes would have to be given up entirely, which of course is not to be thought of. The possibility of obtaining large aeroplanes from America has been considered, but besides the absence of satisfactory design, it would take a very long time for America to become a real producer of aeroplanes, owing to their lack of all modern experience in the matter. In this case, also, is the added difficulty of sea transport.

"(3) The provision of 1000 pilots beyond those already required for the work of the Army and the Navy, could not be hoped for before a somewhat advanced period next year.

"(4) The difficulty of landing a large force of men from aeroplanes in the rear of the Enemy's lines has not been sufficiently considered. If 1000 aeroplanes were to land on a front of 20 miles, it means 50 aeroplanes to the mile, landing where they can on unprepared landing grounds. Such an operation might be possible in certain portions of the desert, or even in certain limited areas on the Continent, but could not be counted on as a general military operation. No doubt something could be done to make the design of the aeroplane suitable for sustaining a smash without killing the passengers, but very few aeroplanes of any known design would survive a crowded journey of this kind.

"I have not entered into the possibility of the success of such an operation provided the passage were safely accomplished, but the prospects appear somewhat doubtful."

When James McCudden had returned to the Home Establishment in February, he already wore the Croix de Guerre and Military Medal he had won as a sergeant, and the Military Cross awarded soon after he became an officer. He spent nearly five months, which included the first Fighting Instructors Course at Central Flying School, teaching new pilots combat technique. The standard of instruction had improved so much, thanks to Smith Barry, that when a recently qualified pilot joined his first squadron in France he had a good chance of shooting down the enemy instead of being an easy victim.

Given a Sopwith Pup, he visited several fighter squadrons to teach them air fighting. In this aircraft he performed his first roll. He had the Synchronised Vickers replaced by a Lewis on the upper wing, which he could tilt up to attack from beneath. One day he went up to intercept an air raid by some twenty Gothas but could not close the range to effective shooting distance. Twenty miles off the Essex coast he was still 500 feet behind, but emptied three drums at the nearest one and smelled incendiary bullets coming past him in return.

The feelings he expressed about seeing the enemy in British air space were exactly the same as those often voiced by Douglas Bader during and after the Second World War. "How insolent these damned Boches did look absolutely lording it in the sky above England. I was absolutely furious to think that the Huns should come over and bomb London and have it practically all their own way. I simply hated the Hun more than ever."*

No. 56 Squadron had been sent home to cope with the Gothas. McCudden went to visit Bowman, who was a member, and met several others whom he knew. He said there was a wonderful spirit, which was entirely different from any other squadron's. On 7th July he went on a three-week refresher course and thence to 66 Squadron in France to obtain up-to-date combat experience before returning to England to continue instructing. Fifty-six had also returned to the Front and invited

* They did not have it their own way in 1939-45. Douglas Bader hated and despised them as McCudden, Mannock, Bishop and other famous fighter pilots did. This hatred of the aggressor seems to be what distinguishes the best from the general run. Pilots in the Second World War were more reticent than their predecessors about speaking their feelings, but have admitted to a similar loathing.

him to dinner. He was impressed by the orchestra that Major Blomfield had assembled, and amused by his method of recruiting them. Conscription had been introduced, so Blomfield visited several London orchestras and asked the names of musicians who were being called up. He then ensured that they joined the RFC and came to his squadron. He would also load half a dozen men of various trades into a lorry, and drive round other squadrons with them. If he found a musician in one of the trades relating to his lorry-load, he would swop the appropriate non-musician for him on the spot and carry him off. Before leaving his hosts, McCudden asked to join the squadron and Blomfield promised to apply for him. On 15th August he duly became one of that hand-picked élite, as a flight commander.

The day of the lone hunter, the fighter pilot dodging in and out of cloud, climbing as high as he could, trying to position himself up-sun from where he expected the enemy to appear, was not quite done and never would be. Bishop, McCudden, Mannock all kept up the practice, as did Fonck, Guynemer and Nungesser; but offensive patrols by formations of up to six machines had become standard. Five was considered the ideal number, either two on each side of the leader or four in a diamond with the fifth above the rearmost man to keep a lookout astern.

These small formations had proliferated. Air fighting had entered the era of what was then called dogfighting, when twenty or more aircraft mixed it in a brawl that spread over a large area of sky and several thousand feet of altitude.*

Hear McCudden on the subject of dogfighting. On the long summer evenings the British had, as a rule, eight fighter formations up in addition to other machines. "The fun used to begin at about 7 p.m." and usually went on until dusk or even later. The enemy normally sent an equal number of fighter formations up. "The evenings were wonderful, as the fighting was very fierce and well contested." At least thirty machines would be weaving about in combat. Now and again one would fall, trailing sparks and smoke, flaming like a meteor until it crashed with an eruption of burning débris and a final gout of flame.

On the 18th McCudden scored his first kill in an SE5: an Albatros. The next day he got another. Two days later he shot down two and on 22nd August, a DFW two-seater. He was well on the way to his next decoration.

During January 1917, Harold Hartney's first month as a flight commander, one of the other flight commanders was killed and the third

*In the Second World War the term "dogfight" was commonly applied to two fighters each trying to get on the other's tail or into some other killing position.

posted: which left Hartney the senior and second-in-command. On 1st February he led a patrol of five FE2ds which was attacked by three Albatroses. Four Fees were shot down, two pilots and an observer were killed, one pilot and two observers taken prisoner. On 14th February Hartney took off on a photographic reconnaissance escorted by another FE2d flown by Lieutenant Taylor. Hartney's observer, W. T. Jourdan, an American who had been a clown in Ringling Brothers' circus, was an excellent gunner and photographer. Seven of Richthofen's Albatroses attacked them. Hartney dived, zoomed, kicked on right rudder and stall-turned 180 degrees into a steep dive. The Germans, instead of breaking and following him down, held formation.

By the time they had wheeled about he was far ahead; with Taylor safely tucked in astern. When the enemy did catch up, Jourdan shot two down and Taylor's observer got a third. Hartney said he knew Richthofen's scavenging habit of picking up cheap kills by keeping out of a fight while his comrades engaged the British or French, then going in to help himself. He turned up now in his all-red machine. A burst from his guns sheared off one of Hartney's propeller blades. The Fee began to vibrate violently. Hartney made a crash landing that knocked him out. He regained consciousness on a duckboard carried by two Australians. Taylor had crashed also and, like Jourdan, was crippled. His gunner had been shot dead. Hartney was sent to hospital and then to England.

He spent some months instructing. He had confirmed his willingness to go to America to instruct, but the American Army found a better use for him. On 21st September 1917 he was ordered to report to Toronto to take command of the 27th American Aero Squadron, with the rank of major in the US Signals Corps. When he arrived in Toronto he found that he had been given American citizenship.

CHAPTER 16

———•———

1917. Equal Poise of Hope and Fear

Airmen's constant hopes and fears were centred on the quality of the aircraft they would be called on to fly, from half-year to half-year, against the enemy. So far, the advantage had fluctuated in overall favour of neither side. For the time being a balance had been struck, but it was always precarious. Even when times were good, pilots and observers were conscious of the transience of aerial superiority. Georges Guynemer could have spoken for any of them when, asked by an admiring lady "What new decoration is there left for you to earn?" he gave the dry rejoinder: "The wooden cross." He was all too prophetic. L'Aviation Militaire suffered as badly as the RFC under the Albatros's twin guns and from its speed and manoeuvrability.

The demoralisation of the dispirited French land forces, mentally and physically battered by the defence of Verdun, cast shame on the air Service by association. General Nivelle's reckless offensive in April 1917 had caused wholesale desertions and worse; mutiny in sixteen corps. Pétain, appointed first to be Chief of the General Staff and then to replace Nivelle, quelled the mutinies within a month. But it cannot have been easy for the air force to maintain its pride and the scale of its flying effort when another arm refused to go into attack.

By May, Guynemer was the longest-serving member of The Storks and still one of the youngest. On one day, 25th May, he shot down four of the enemy. On 11th June, with his score at forty-five, he was made an officer of the Legion of Honour. By the end of August he had fifty confirmed victories and was the leading ace. Les Cigognes were now flying the Spad 13, whose 220-h.p. Hispano-Suiza engine gave it a maximum speed of 132 m.p.h. at 10,000 feet, to which it climbed in 8 minutes, and a ceiling of 24,000 feet. It was armed with twin Vickers synchronised with the propeller.

This aircraft, the Sopwith Triplane and the SE5 were the ones that took air superiority away from the Albatros DIII and DV. Guynemer, with fifty-four kills, returned from three days' leave in September to find his personal Spad unserviceable. Instead of taking over someone else's, he flew whichever was available. This was invariably one that was awaiting the arrival of a new pilot – who, as in the RFC, would be given the most work-worn – or, even worse, due for overhaul. In one day three of these tatterdemalion ruins let him down with engine trouble or a structural defect. He made three forced landings that would have been the end of a less able pilot. In two fights, his guns jammed.

On four successive days he flew four two-and-a-half-hour sorties. These absurdly long periods at maximum alertness, added to the constant misbehaviour of his aircraft or weapons, brought about a mental condition that matched his physical appearance in a photograph taken at that time. Sunken-eyed, emaciated, he looks worn out and a little demented. He was tortured by the thought that his aborted flights and lack of combat success would make him suspected of cowardice. Insomnia kept him on his feet at night. Often he would rouse his mechanics to run the engine of his current Spad, to be sure it would be fit for take-off at dawn. He was awaiting the arrival of a brand-new one from the factory.

On 11th September, under an overcast sky and in the rain, he took off at 8.30 a.m. with Sous Lieutenant Bozon-Verduraz and found a solitary two-seater, which they attacked. It was bait for an ambush by three Albatroses, which dived from the cloud fringe. Guynemer's No. 2 spotted them, turned to take them on and got away after the skirmish. He did not see what happened to his leader.

Some days passed before the Germans revealed that Leutnant Kurt Wissemann had shot Guynemer down. Neither his aircraft nor his body has ever been found.

Nungesser had returned to Les Cigognes in May after a long spell in hospital having sundry bones broken and reset. The enemy instantly noticed his ghoulishly decorated Spad in the sky and on 12th May an Albatros dropped a message challenging him to single combat. That afternoon, when he arrived over Douai, there were six Albatroses treacherously awaiting him. He shot down two and the cowardly remainder retired.

Nungessser was trying hard to overtake Fonck's growing total, but before he could do so he crashed his Mors car, killing Pochon, his mechanic, breaking his own jaw and sustaining other injuries. He replaced the Mors with a white Rolls-Royce tourer.

In the many ways in which Fonck, Guynemer and Nungesser differed from one another, it was the cold precision of the manner in which Fonck did his work that best exemplified his character. In comparison, Guynemer was often profligate with time and ammunition, and Nungesser totally wild, as ferocious, brave and regardless as a fighting bull. Fonck used to tell new pilots that he found ten, or, at the most, fifteen rounds enough, whereas Guynemer would fight for half an hour or more and Nungesser would thunder in blindly, intent on hitting his opponent and careless of return fire.

Another of Fonck's dicta was that if a bullet were ever to hit him he would apply for a transfer to the trenches. Wounds, in his view, meant lack of skill. He justified this by going through the war unscratched, with the highest score of any Allied pilot. Although his victories totalled five less than Richthofen's, they were worth a great deal more. It is said that not only did Richthofen prey on the easiest targets, such as the BE2c, but also when he attacked in company with others he often merely finished off an aeroplane they had damaged badly, then claimed the kill.

Fonck also exercised a certain degree of caution, which he would probably have called prudence: unlike Nungesser, who was more charging lion, rampaging elephant or wild buffalo at full tilt than game hunter, or Guynemer, who also thought less of self-preservation than Fonck. Fonck's warning to inexperienced pilots was that it was less dangerous to attack fifteen enemy aircraft on the Allied side of the lines than five on the German side. Not that he was timid about being outnumbered. He said that a large number of opponents got in each other's way and often held their fire for fear of hitting one of their own side. When he came upon a formation he tried to shoot down the leader, then, in the subsequent confusion, a second.

He had devised a collimator – an instrument for adjusting the line of sight – to make his shooting as accurate as possible. Mounted between the guns of his Spad, it was a tube with a plain glass lens on which were two concentric circles. One covered a field of one metre, the other of ten, at a range of 100 metres. Knowing the wingspan and fuselage length of every enemy aircraft, he calculated its precise distance from the aspect it presented.

Fonck has been criticised for conceit, aloofness and a coldly calculating approach to air fighting. He might have been a trifle conceited. Aloofness is no offence: it is usually the product of shyness or a quiet disposition, not a sense of superiority. As for the third complaint of his detractors: they were jealous of his infinite capacity for taking pains. His admirable traits are seldom mentioned. He was decent, self-respecting and loyal. Opinion about him was so fiercely divided among The Storks that it threatened to disrupt the harmony essential to an efficient unit. Colonel Duval, who was the latest of Barès's successors, offered him a posting, which he refused. He was then offered command of the escadrille and the riddance of those who differed from him in his views on air fighting. He asked what would be done with his Commanding Officer. The answer was that he would be shifted. Fonck angrily dismissed this as a dirty trick and said he would never accept promotion at another's expense. Embittered by intrigue, he withdrew further into his own company and went hunting alone.

If Barès or du Peuty had still been "Responsable de l'aéronautique" at General Headquarters, the matter would have been handled with far greater finesse and no hint of intrigue. But Commandant Guillabert had replaced Barès on 15th February 1917, and was himself followed at the end of April by Commandant du Peuty: who left on 3rd August to join an infantry regiment – despite being an ex-cavalryman – and was killed. It was Duval who took his place and held it until the end of the war; by when he had become a general.

Fonck is often credited with being the only pilot on either side to shoot down six enemy aircraft in one day. In fact, Captain J. L. Trollope and Captain H. W. Woollett, both of 43 Squadron, did it with Sopwith Camels in March and April 1917 respectively.

We have to leap forward in time to the day on which Fonck scored his greatest number of victories, but this is the appropriate and logical place for the story. He had declared an intention to bring down five hostiles in one day; which prompted more argument and the laying of bets. On the morning of 9th May 1918 he was called at four o'clock. It was misty, but he took off; only to return after half an hour. He went back to bed, telling his batman to call him at noon. The mist was still thick then, so he went to lunch. At 3.30 p.m. the sun came out and by 3.45 had dispelled the mist. Fonck took off with two others. More than half an hour later he saw a Rumpler escorted by two Halberstadts. As instructed, his companions held back. He dived and his bullets severed the wing of one Halberstadt and set the other on fire. The Rumpler turned for home. He chased it and sent it down in flames. It was 4.20 p.m.

He stayed for a while, then returned to base to refuel and rearm. He was called to the telephone to receive the congratulations of General

Debeney: the kills had been reported by the men in the trenches. As soon as possible he took off once more. Capitaine Horment, his CO, sent two others to keep an eye on him. He found an Albatros and shot it down at 5.17. Then he picked out several specks in the sky that had to be enemy fighters. On approaching he identified them as five Fokker DVIIs in V formation, with four Pfalzes in a diamond well beneath them. He stalked them through cloud and picked off the rear Pfalz; then closed to within forty metres of the Fokkers without being seen; and shot down the leader with just four rounds.*

In his autobiography Fonck wrote that to obtain good results a fighter pilot must know how to control his nerves, how to have absolute self-mastery and how to think coolly in difficult situations. "I always believed that it is indispensable to maintain absolute confidence in ultimate success, along with the most complete disdain for danger." That last holds the secret of what separates the supreme champion from the also-rans; even the closest of his contenders. The other three precepts are mere glimpses of the obvious. To maintain absolute certainty of victory, to refuse to contemplate failure: there lies the touchstone in any field of endeavour. In air fighting there is an added element which makes it different from any other: it is one's life that is at stake, not fame or wealth.

The second crisis of 1917 was the confrontation of the RFC by the first of the newly formed Jagdgeschwader, commanded by Richthofen and comprising Jastas 4, 6, 10 and 11. Each of these had two flights of six aircraft. Operating much the same as a RFC Wing, the Jagdgeschwader were differently used in one significant way: they were tactically mobile and frequently moved about the Front at short notice.

General von Höppner recorded: "Because of their number and their sporting audacity, the English continued to be our most dangerous adversaries and, as before, the major part of the German air strength was concentrated against them." The Jagdgeschwader were also a considerable menace to the French. Nos 2, 3 and 4 Jagdgeschwader were not formed until 1918, but the threat posed by No. 1 was especially unsettling because the RFC was still reeling under the effects of April.

With the birth of the Jagdgeschwader, and putting it in the hands of

---

* Shooting down an aeroplane with four bullets might sound incredible, but Squadron Leader "Ginger" Lacey did the equivalent in 1945 over Burma, when he shot down a Japanese Nakajima Ki43 ("Oscar") fighter with a half-second burst from the two 20-millimetre cannon of his Spitfire VIII: nine rounds. Allowing for the disparity in rate of fire between a machinegun and a cannon, and the range – Lacey fired at 200 yards – the comparison is close.

Richthofen, came a display typical of primitive tribal mentality. Having first had his own aircraft painted scarlet all over, he now ordained a vulgar miscellany of colours for the whole Geschwader. This was, psychologically, equivalent to the behaviour of savages who painted their faces and bodies to terrify their enemies. To the RFC it was a mere display of boastful arrogance in the worst taste. They dismissed No. 1 Geschwader with derision as "the Flying Circus".

One of Richthofen's pilots was Ernst Udet, who was to amass a score of sixty-two kills, but in 1917 was a novice fighter pilot who met Guynemer in combat. He was airborne at 5000 metres when from the west an aircraft approached. "It quickly grew in size and I recognised it as a Spad." They met at the same altitude. "I saw the other man's machine was painted light brown. Soon we were circling round each other playing for an opening. The first man to go behind the other's back was the winner. Sometimes we passed so near to each other that I could see every detail of my opponent's face: that is, all that was visible of it below his helmet." On the Spad's side were a stork and two words painted in white: "Vieux Charles". He knew now whom his adversary was.

"I knew that I was in for the fight of my life." After several minutes' manoeuvring, "for a moment I had him in my sights. I pressed the trigger ... there was no response ... my guns had jammed!" Holding the stick with his left hand he pounded the guns with his right fist. It was unavailing. He considered diving away, but knew that Guynemer would be on his tail in a flash. They continued circling. "It was a wonderful flying experience; if one could forget that one's life was at stake. For a while I completely forgot that he was Guynemer, my enemy. Rather it seemed that I was having some practice with an old friend." Suddenly Guynemer looped and flew on his back over Udet's head. Udet hammered again at his guns. Guynemer saw this and "knew I was his helpless victim".

Guynemer again passed close overhead. "And then to my great surprise, he raised his arm and waved to me. Immediately afterwards he dived away towards the west. I flew back home, stupefied." And lived to kill many more Frenchmen and Britons. Richthofen would never have spared an enemy; which was acceptable, even if his guns jammed. Guynemer's action was one of genuine chivalry. Udet was a likeable man and, in later years, no admirer of Nazism. He was a pilots' pilot, brilliant in all aspects of his craft. At heart he was always a dashing, roystering fighter boy. He said, after his encounter with Guynemer, that many of his comrades attributed the Frenchman's restraint to the jamming of his own guns; but he himself insisted that it was pure chivalry that had prompted Guynemer to spare his life. Udet was plainly no cynic.

Richthofen, inaccurately known to the British as "The Red Baron", was more fittingly dubbed by his own countrymen "The Red Devil". The spring and summer of this year gave him some easy pickings; but he came within a millimetre of being cut off in his prime when on 6th July he led Jasta 11 in an attack on twelve FE2s and RE8s, escorted by Collishaw's flight of Triplanes. He himself, naturally, went for a Fee – the Tripes would have been far too lethal – and got a bullet across his scalp for his pains. The wound knocked him out, he recovered his senses, made a hasty landing, fainted again, then spent a month in hospital.

Voss, who was then second to Richthofen in victories, commanded Jasta 10. Despite his quiet modest nature, even he was infected by the Flying Circus's ostentation. His Fokker Triplane was light green, with a white ring. The wing under surfaces and wheel discs were pale blue. A threatening face was painted on the prow, with the engine air intakes as eyes.

This aeroplane was first seen at the Front in July. Like the first Fokker fighter, it was not the product of its designer's own ingenuity. This time it was not a copy of the Sopwith, but inspired by it. The Sopwith Triplane had astounded the Germans. In one skirmish, a solitary Tripehound had out-manoeuvred eleven Albatros DIIIs. The German Air Ministry asked industry to produce a triplane. Anthony Fokker frequently visited the Jastas, so had heard all about the Sopwith from Richthofen, and his Chief Designer, Platz, was already at work on a design. Two of the first to go into service lost their wings in a dive. All were grounded until modifications were made. The best speed was 102 m.p.h. – 15 slower than the Sopwith – at 13,000 feet and the armament was two synchronised guns.

The Schlachtstaffeln – Schlastas – were being equipped with purpose-built ground-strafing types whose crews were better protected than the Allies who had to do this work. The Hannover CLIIIa two-seater weighed 2378 pounds, was armour-plated underneath, armed with a Spandau for the pilot and a Parabellum for the observer, and capable of 103 m.p.h. The Junkers JI was similarly armoured and had a corrugated metal skin. It carried two Spandaus and a Parabellum. It weighed 4787 pounds, so its 200-h.p. engine gave it only 97 m.p.h.; but its strength and firepower were what mattered, not speed. The Halberstadt CLII was another machine with an armoured belly and a speed of 97 m.p.h. It had one Spandau and one Parabellum. Weighing 2532 pounds, it was most manoeuvrable and able to evade ground fire.

Fritz Ritter von Röth, who had been so gravely wounded in 1914 that he was nearly invalided from the Army when discharged from hospital in 1915, but managed to get himself to pilots' school, had been flying in a reconnaissance Staffel. On 31st March 1916 his aeroplane had plunged into the ground from 150 metres when its engine cut. The archives' bland comment is: "It was a wonder that he emerged unscathed from the wreckage." Feeling that he had no more bad luck to come after what had happened to him in the first weeks of the war and early in his flying career, he did not relinquish his hopes of a posting to a fighter Staffel. His wish was fulfilled in February 1917, but he found that not all fighter pilots were given the opportunity to shine as brightly as the ornaments of Jasta 2 or Jasta 11. He had to wait a long time before success came to him. It was not until 26th January 1918 that he was mentioned in despatches, when he scored his third victory in combat. But it was as a destroyer of balloons that he won fame later in the war.

His peer as a balloon-buster, Heini Gontermann, made a first unexpected encounter with the enemy while still under training. He was in the final stage of his course and flying a twin-boom Ago when he unexpectedly met a French Caudron. Closing to seventy-five metres, he "let his machinegun speak for him", according to the records, and surprised the Frenchman, who took evasive action; but too late. Gontermann joined a unit equipped with the Roland Walfisch – Whale – described as "that clumsy but good aeroplane", a reconnaissance two-seater with only a Parabellum for the observer. Gontermann praised it: "It climbs like a monkey, flies like the Devil and is very manoeuvrable." He had it painted in many-coloured stripes, green predominating, and the troops at the Front far and wide used to talk of this "coat of many colours".

In spring 1916 he was sent to a single-engine fighter school. Speaking of the tensions that awaited him as a fighter pilot, he wrote in a letter home: "My body is in my hands, my soul in the hands of God. That gives courage. And always to keep this undaunted will be my ideal." Posted to Jasta 5, he shot down his first opponent, "an Englishman", five days later, in flames, on his first patrol.

On Good Friday 1917 he brought down his seventh, another Briton. In a dogfight, the whole Jasta against an RFC squadron, the CO collided with another Albatros and was killed. Gontermann was given command. On Easter Sunday afternoon he was on patrol with three companions. Observation balloons were aloft. Gontermann himself described this first balloon attack and it is interesting that it set the pattern for his subsequent success as balloon-buster.

"Wind and cloud were favourable. I waited a while for the most opportune moment. There was no enemy in the vicinity. I was 3000 metres over our lines. In a quarter of a minute I reached the balloon and attacked it four times with incendiary bullets. It caught fire on the fourth, as I was about to abandon hope. Anti-aircraft fire opened up at me as I made for home. The same afternoon at about six o'clock I took off for another attack. On my third run my guns jammed. The job seemed too risky. Anti-aircraft shells were bursting. The balloon observer made a parachute jump. I returned home through flak but unharmed. Unfortunately the balloon did not burn. I think both balloons were made of some special flame-resistant material. I will try with a home-made incendiary dart. I hope that will be quicker and more certain."

On 30th April he was transferred to command Jasta 15, where he received a letter from the General Commanding the Luftstreitkräfte on 5th May, confirming his fifteenth victory – both balloons and aircraft – at 11.50 a.m. on 26th April, near Arras. "By his attacks against balloons and opponents he has demonstrated outstanding keenness, perseverance and intrepidity in most difficult circumstances. I send Leutnant Gontermann my very special appreciation. Von Höppner."

Between 6th April and 11th May he raised his total from five to twenty-one. On 15th May he was decorated with the Pour le Mérite.

After a long leave he returned to the Front refreshed, "to embellish his successful and victorious career". On 18th August he brought down four balloons and an enemy aircraft, and by the next day he had accounted for seventeen balloons and seventeen aeroplanes in all.

Jasta 15 had acquired the Fokker Triplane and Gontermann flew it at every possible moment. He said he hoped that "with these we shall have greater success than the Richthofen Staffel".

He took off to practise one day and quickly climbed high above the watchers on and near the airfield. "It was for us all a huge enjoyment," wrote one eyewitness, "to see him gliding about in the bright sunlight and follow his manoeuvres with our eyes. Suddenly we saw him loop and dive steeply. As a layman, it seemed to me at first sight that he was carrying out one of the stunts that the most outstanding flyers perform. A cry of terror from members of the Staffel standing near me told me it was something more serious." The machine crashed to the ground. The onlookers rushed to it and found Gontermann unconscious and covered in blood. The Medical Officer gave him first aid. He was rushed to hospital for an operation. He died that evening: twenty-one years old with forty victories to his name.

Carl Degelow, when he joined Jasta 36 in August 1917, found them equipped with Albatros DIIIs which were being replaced by Fokker Triplanes. The CO was absent and his deputy, a haughty Prussian, was in command. He sent Degelow up to practise air-to-ground firing.

Degelow learned later – too late – that he had been told to make a low pass and line up the targets, without firing, then to come in for a second run, shooting. Engine noise prevented him from hearing these instructions. On the first pass he had the targets perfectly lined up, so thought to impress his new comrades and opened fire. Climbing away, turning, he saw a great commotion on the ground: men running about on the range, others shooting off flares, signalling him to hold his fire. He made a low, slow pass and saw someone being carried away on a stretcher. He thought he had killed him. When he landed, he found an officer who had been behind one of the targets lying on the grass pale and still while a medical orderly bandaged one of his feet, wounded by Degelow.

The Second-in-Command made a summary decision. "Pack your bags, Herr Leutnant, you're finished here." No use arguing. "Befehl ist befehl ... Orders are orders, the Prussians' holy words," was Degelow's observation. He had to linger for three embarrassing days while his posting was arranged. On a Friday he departed in the Mercedes that had delivered him the previous Monday. He felt that he was finished as a pilot: "Who would want such an impetuous fool in his unit?" But he was merely sent back to the single-seater school.

Thence he went to fly Pfalz DIIIs for Jasta 7, whose CO was the popular Leutnant Josef Jacobs, who said: "We will discuss the incident at Jasta 36 just once. Agreed?" Only too readily, thought Degelow, relieved.

On 2nd September he took off with seven others. The Pfalz was a pretty aeroplane, with a slimmer and more tapered torpedo-shaped fuselage than the Albatros, carrying two synchronised Spandaus and capable of 102 m.p.h. At 3000 metres they saw 25 British aeroplanes on their way to bomb the main arsenal at Ghent. Degelow described them as Bristol Fighters, but they might have been DH4s. He attacked the rearmost one, used all his ammunition, but did not fetch it down. He could hit brother officers on the ground, but apparently found moving aircraft à bit too much. The Brisfits turned to fight and one of them punctured Degelow's oil tank. He had to break off because his goggles were covered in oil.

The next day the Jasta went up to intercept a formation of Brisfits, Camels and Spads. Five of the German pilots, including Degelow, each took out a Camel. That was a considerable tribute to their skill as much as to their aeroplane. One of the Bristols was flown by a New Zealander, Captain Keith Park.*

No. 3 Squadron RNAS, commanded by Flight Commander F. C. Armstrong, DSO, often met Jasta 7, who called it "The Armstrong Boarding School", because to gain experience it was "assigned to various sectors of the Front and indoctrinated in the manner of an English boarding school. They were always welcome adversaries: their lack of experience made them easy prey," Degelow gloated. This is sheer calumny: Naval Three never comprised only novices.

When Degelow forced down one of them, a twenty-year-old Canadian, Flight Sub-Lieutenant H. S. G. Youens, the Jasta entertained him in the mess, which was customary. After a bibulous dinner, Youens mentioned that he was a violinist. A fiddle was borrowed from a mechanic, but had no A string. Youens produced two complete sets of strings and, accompanied by a piano, played for his captors. Degelow recalled: "As the opening piece we played 'Deutschland über Alles' and to the delight of all present Youens played the song with as much intensity of feeling as though it were 'God Save The King'. The whole evening he gave us great pleasure. We eventually did play the British national anthem and every German stood at respectful attention as a sign of comradeship beyond the bounds of national or political affiliation. Thus a defeat was transformed into a victory." But they put their prisoner to bed in a room with boarded windows and took away his boots and braces.

On 23rd May 1917, after ten hours' intense artillery bombardment, shortly before the Italian infantry sprang from their trenches, a wave of thirty-four Caproni with fighter escort dropped tons of high explosive on the Austrian front line and reinforcement assembly area. A second wave followed, consisting of forty-two assorted smaller bombers and thirty Nieuports. This first mass participation at low level in support of the ground troops, within artillery range, was greatly extolled. Material results were, however, limited by inaccuracy and wide dissemination of the bombing; but the effect on both the Italians' and the enemy's morale was considerable.

The Italians claimed that the event "enraptured the souls" of their troops, and terrified the Austrians "all the more because the unexpected incursion found their spirit unprepared to endure it".

D'Annunzio had done his bit by writing an exhortation to the airmen

---

* Who, as an air vice marshal, commanded No. 11 Group of Fighter Command in the Battle of Britain. This was the Group guarding the hardest pressed area: south-east England. He retired as an air chief marshal with a knighthood. It gives pleasure to record that in this dogfight, when an Albatros attacked him, he hit it in the petrol tank and it exploded.

before the battle. It opened: "Italian aviators, winged guardians of our sky, aerial precursors of our armies on the ground ..." and went on to praise their achievements to date "in bringing constant travail to the foe, in providing with obstinate audacity for the paucity of equipment and the reverses of fortune". It was skilfully couched in terms to which his countrymen would respond.

The General Staff, however, declared the first aerial operation by large numbers praiseworthy in intention but confused in execution: the sum of too many individual actions rather than a harmonious integration of them all.

D'Annunzio did more than try to incite the Air Corps to maximum endeavour, he flew on the massed bombing raid on 23rd May. For this he was awarded his third Silver Medal for "his vibrant and convincing words, his ardent example". On 29th September he was promoted to major.

On 10th June, in another big attack, 32 Caproni, 56 reconnaissance types and 53 fighters took off. On account of very bad weather, 20 bombers returned without having reached their targets. The rest dropped their loads blindly through cloud. On the 19th 145 aeroplanes took part in the battle.

The Eleventh Battle of the Isonzo opened on 18th August 1917 on a front of forty-two kilometres with the usual artillery blast. Spotting aircraft were airborne all day. On the 19th, when the infantry went in, 85 Caproni bombed the enemy rear, reconnaissance aircraft dropped small bombs and ground-strafed, and in all 280 aeroplanes operated that day and 261 the next. Between the 19th and 28th August there was an average of 225 serviceable aircraft every day and 1474 sorties were flown. Of 300 pilots and observers involved, 81 were killed.

September began with an act of great heroism. Sergeant Arturo Dell'Oro, on solo fighter patrol, attacked a two-seater. His gun jammed. Rather than abandon the fight, he rammed the enemy aircraft's wing with his undercarriage. Entangled, both crashed. All three occupants were killed. Sergeant Dell'Oro was posthumously awarded the Gold Medal.

Baracca was always the leading Italian fighter pilot. By the first week of June 1917 he had scored thirteen confirmed victories, been promoted to captain and was commanding the 91st ("The Aces") Squadron, flying Spads. With him were Lieutenants Barrachini and Prince Ruffo di Calabria. On 31st July Baracca brought down his fifteenth victim with only four bullets, fired from thirty metres. He constantly risked wing-to-wing collision with enemy aircraft in his endeavours to get on their tails at close range.

Ruffo di Calabria had shown himself a born fighter pilot in his first combat, flying a Nieuport 11, on 23rd September 1916. On 1st January 1917 he shot down his fourth Austrian machine. In February he had five fights, in four of which his opponents quit. The fifth, Ruffo abandoned with a jammed gun. In April and May he was in twenty-six combats and made four kills.

Other distinguished members of the 91st were Ferruccio Ranza, who eventually scored seventeen, Luigi Olivaro who brought down twelve, Gastone Novelli, who finished with eight, and three who made six kills: Cesare Magistrini, Bortolo Costantini, Guido Nardini.

Thirty-seven-year-old Pier Ruggero Piccio, in June 1917 a major and commanding First Fighter Group, had six victories to his name. Between July and November he scored eleven more, two of them in one action. On 19th August, after leading a fighter escort for a Caproni squadron, he refuelled and went back over the lines alone. He soon saw an enemy aeroplane flying high towards the south-west and climbed above it. He dived, approached, aimed; and was about to shoot, when he was amazed to see the enemy burst into flames and fall, smothered in fiery smoke. Looking around, he turned his aircraft and saw Francesco Baracca: who had mischievously poached his target from him with one short burst.

Disaster befell the ground forces on 24th October, when Austria–Hungary launched a well-prepared offensive. The Italian generals' ineptitude had already led to a succession of defeats and heavy losses. This time, their troops broke and the enemy swept across sixty miles to the Piave River in eighteen days, taking a quarter of a million prisoners.

The British and French each rushed an army corps to the Italian Front. The RFC moved two squadrons there. Four more followed soon after.

No American Expeditionary Force or air squadron had arrived in France yet, but as early as March 1917 a conference had been held in Rome to agree terms for the training of American pilots in Italy. It was agreed that 500 pupils would be accepted in the flying schools at Foggia South and Foggia North, with a further 500 to come later. The first course began on 28th November 1917.

CHAPTER 17

———————•———————

1917. The Allies Resurgent

"He was a tiny little fellow and used to have blocks of wood fixed to his rudder bar pedals so that he could reach them," said his first flight commander, Captain K. G. Locke. "A well educated and keen young officer always willing to learn."

Sholto Douglas, his squadron commander, said: "He was courageous, a first rate shot and a good formation leader. His moral effect on the squadron was excellent. His eyesight was exceptionally keen and he could spot any enemy aircraft further away than anyone else."

They were talking about Andrew Weatherby Beauchamp-Proctor, who was born on 4th September 1894, in South Africa; stood only five feet and one inch tall; destroyed 22 enemy aircraft, drove 16 down and shot down 16 balloons; and won the VC, DSO, MC and bar, DFC, Croix de Guerre and Légion d'Honneur.*

"A quiet and likeable chap," said one squadron comrade. Another described him as: "A very likeable chap. Not a particularly good pilot, but an excellent shot." Someone else: "He would attack anything in the sky, oblivious to danger."

"'Procky' was a very serious-minded young fellow," a colleague recalled. "As a fighter pilot he had one great advantage: twenty-five

---

* He was not the only short pilot to gain fame. There were many in both world wars who were less than five feet six inches tall: Richthofen among them. The smallest – and little known – RFC or RAF aircrew was V. C. (his initials, not the decoration) "Shorty" Keough, one of the many Americans with a civil flying licence who came to England in 1939 and 1940 to volunteer. "Shorty" was four feet ten inches and nearly failed his medical, but demonstrated that, seated on two cushions, he could see out of a cockpit. He flew Spitfires in the Battle of Britain and was killed a year later.

pounds less weight made a great difference in the performance of the SE5A."

"Procky" Beauchamp-Proctor had been an engineering student at South African College, later the University of Cape Town, before joining the Duke of Edinburgh's Own Rifles as a signaller on 1st October 1914. He served in the South-West African campaign, transferred to the South African Field Telegraph and Postal Corps, and was demobilised in August 1915, whereupon he resumed his engineering studies.

Early in 1917 he met Major A. M. Miller, DSO, RFC, who was on his second recruiting tour of South Africa, volunteered for the Flying Corps, and was attested as a third-class air mechanic on 12th March. He began training at South Farnborough on 13th April and after six weeks went on to No. 5 Reserve Squadron at Castle Bromwich. Pupil pilots had a long working day and made short flights. On 25th May he went up at 5.20 p.m. for 15 minutes' dual in a Farman Longhorn at 500 feet. At 8.30 p.m. he was airborne for half an hour's dual at 3000 feet. On 2nd June he flew from 6.40 to 6.50 a.m. and from 6.55 to 7.10 a.m. These were typical days. On 10th June he was up from 4.40 to 5.15 a.m., before making his first solo, 5.30–5.55 a.m., at 1500 feet. He crashed on landing and broke his undercarriage and left lower mainplane. By 16th June he had accumulated 9 hours 50 minutes, 4 hours 35 being solo. He was given his wings on 29th July. He joined Sholto Douglas's 84 Squadron, where he soon crashed again. By then he had 33 hours 20 minutes flying time, 20.15 solo. On 21st September the squadron moved to France and on 14th October he had his third crash when his engine seized. He flew nineteen operations between 21st October and 22nd November, helped to bring down a balloon and crashed on landing.

Up on an engine test over Arras on 5th December he met a two-seater, fired both guns from 200 yards and sent it into a vertical dive. He could not follow it down, because it had damaged his tail plane and control wires with return fire. He was awarded a probable.

On 15th December he crashed an SE5, landing with no petrol, and wrote off the radiator, propeller and upper planes.

A few days later he received news that both his parents had died, within a few days of each other. This, his comrades said, changed his attitude from one of placidity to extreme aggression. He made his first kill shortly after. But that was in January.

1917 was a vintage year for aeroplanes as well as for pilots. The Germans introduced the Albatros DV and the Fokker Triplane. The former was a V-strutter and prone to flutter and collapse of the lower wing in a dive, but fast and handy. The three German ground-attack fighters pioneered this specialist type.

Among the French the Spad XIII was even better than the VII, and, when fitted with a Rolls-Royce engine capable of 133 m.p.h. and a climb to 10,000 feet in 8 minutes. The Nieuport 28 was almost as fast. The Breguet 14B2 was an excellent reconnaissance and bomber aeroplane, strong and reliable. The ground-attack variant had two Lewis guns, front and rear, inclined downward, and the crew were protected by armour plate. The Voisin VIII night bomber had a quick-firing 37-mm Hotchkiss cannon, a formidable piece of ordnance.

But the British had the best of the year's innovations. The F2B Bristol Fighter is usually rated the best all-round performer of the war. The Rolls-Royce-engined SE5a could fly at 130 m.p.h. The Sopwith Camel was a devastating fighter, the first British aeroplane fitted with twin synchronised Vickers guns. The Sopwith Dolphin was the war's most under-rated fighter, with excellent performance at high altitude and a top speed of 128 m.p.h. It also had two synchronised Vickers, and one, sometimes two, free Lewis guns. The DH5 was a fine single-seater ground-attack type. The DH4 was the best day bomber of the war. It carried two 230-pound or four 112-pound bombs and was armed with one, later two, Vickers for the pilot and a Lewis for the observer. The Handley Page 0/100 and 0/400 bombers were gigantic: wingspan 100 feet, length almost 63 feet. Their two 250-h.p. Rolls-Royce engines gave a speed of just under 100 m.p.h. Both carried a crew of four, the 400 sometimes five, and five machineguns. Their bombload was 2000 pounds.

The Camel, officially the Sopwith F1, derived its name from the hump over its two guns, but its performance in no way resembled the ambling gait of its namesake. At 118 m.p.h., its top speed was not exceptional, but it reached 10,000 feet in 10½ minutes. Its handiest trait was the astonishing speed with which it turned to the right, thanks to the tremendous torque of its 130-h.p. Clerget rotary engine. It could flick right round in a full circle in the time it took any other fighter to turn 90 degrees. It was so quick that a pilot wanting to face an adversary on his left could turn right-handed through 270 degrees faster than he could turn 90 degrees to the left. It was a handful to control and its violent turn easily became a spin. Flown by an experienced pilot it was a superb weapon. More enemy aircraft were shot down with this fighter than with any other on either side.

These were the aircraft that contested the air battles of the last seventeen months of the war: closely matched, many of them flown by pilots highly experienced in combat, well armed and with ample endurance, they were able to fight in a manner that had not been possible for their predecessors. Their pilots still had to labour for breath in the rarefied air at high altitude, although crude methods of oxygen supply were gradually being developed. They still suffered from agonising cold. Those who flew behind rotary engines still had to inhale the sickening fumes of castor oil and find their view obscured by the stuff, which smeared goggles and windscreens and dirtied faces. Guns still jammed.

On 31st July 1917 the Third Battle of Ypres began. This is the long battle that is usually referred to as "Passchendaele", the ridge seven and a half miles north-east of Ypres that was the last objective won; on 6th November.

Air activity increased during the three weeks preceding the battle – by which time Richthofen was back, head still bandaged – until the first great dogfight of 26th April when 94 aeroplanes mixed it at various levels from 5000 feet to 17,000 feet. At the bottom of the stack a few unspecified German two-seaters awaited a chance to reconnoitre behind the British lines. At 8000 feet 30 Albatros DIIIs and DVs and Halberstadts attacked 7 DH5s. Between 12,000 feet and 14,000 feet, 30 Camels and SE5s fought 10 Albatroses. Another 10 Albatroses and 7 RNAS Sopwith Triplanes were engaged at 17,000 feet.

The next day, 8 FE2ds entered the same area to tempt the enemy: who obliged by sending 30 Albatroses and Halberstadts to try to take them out. Then 59 SE5s and 5As and Sopwith Triplanes hit them. One SE5 was lost and nine of the enemy were brought down.

Arch Whitehouse, the long-suffering aerial gunner, claimed to have started the first real dogfight over the Western Front. By the time 22 Squadron exchanged their FE2s for Bristol Fighters he was credited with seven enemy aeroplanes and six balloons. He could also, he said, fly the Fee so well, after being given dual instruction by several pilots, that Captain Clement, with whom he usually flew, tried to arrange for him to remuster to pilot on the spot without going to flying school. He should have been thankful for his prospects of living to see the end of the war, instead of resentful, that the request was turned down. He alleged that it was because it would have offended "the caste system", which was ridiculous. It would have offended good sense: to be a real pilot he had a great deal more to learn.

He had recently been awarded a Military Medal when he set off with Lieutenant Youl, deputy leader of a formation of six Brisfits, to escort twelve DH4s that were going to bomb Gontrode, forty miles behind the enemy lines. Soon after the rendezvous the engine began spluttering, so he stood to look over the pilot's shoulder at the instruments, decided the fuel line was blocked, and gave the pressure pump a few strokes.

(Most pilots would have reprimanded him for interference.) Then he "told" Youl to try to catch the others up, as the engine note improved. The engine faltered again. Whitehouse advised Youl to turn back, which he did. The engine continued misbehaving. So, according to Whitehouse, he told Youl to change tanks. The rest of their formation were long gone. The engine resumed running smoothly, so Youl turned towards the target once more. Six Belgian Spads joined them. Soon a flight of SE5s tacked itself onto the group that was following Youl. Presently another flight of Spads and some Camels added themselves to the formation.

One of the Camels fired a red Vérey light. A large German formation was in sight overhead. Whitehouse told Youl to climb (he claimed). Finally, according to him, he fired a signal and "led the pack of at least thirty fighters" into the fight. Lieutenant Youl seems to have been a remarkably docile and obedient officer, to take his orders from a corporal. There was pandemonium. Tracer was zipping about in all directions, machines were colliding, bits of aeroplane were fluttering down, men were tumbling from their shattered cockpits. He saw an SE5 pilot climb onto the wing of his burning machine and, reaching into the cockpit, try to sideslip it to the ground and direct the flames away from himself. The aeroplane exploded. This feat was not unique. Eyewitnesses in other combats have described pilots resorting to this desperate measure.

Two days later Whitehouse crossed the airfield to look at 56 Squadron's SE5s. McCudden spoke to him. "When are they sending you home?" (For pilot training.) "I don't know, sir. I expected to be sent home after fifty hours over the lines." "I hear you've done very well. How long have you been flying?" "I've done nearly four hundred hours as a gunner."

That was a huge amount of operational flying and Whitehouse deserved more than one MM for it. But he did eventually go back to England and become a pilot, after having been put up for a Distinguished Conduct Medal.

Third Ypres had begun with the Germans at a disadvantage. Their total strength was 600 aircraft, of which 200 were fighters. The Allies had 320 fighters out of a total of 852. Throughout the three and a half months of the battle, ground strafing went on in increasing strength and bombing continued day and night against tactical targets near the front and strategical ones deep in enemy terrain. A few Camels sometimes flew at night, trying to intercept the German bombers that were returning the attention their side was receiving from the British and French nocturnal raiders.

The arrival of the RFC in Italy brought to prominence the pilot whom Billy Bishop, after the war, called "the finest fighter pilot the world has ever known". At that date, most people would have awarded this honour to Bishop himself. Bishop's nominee was another Canadian, William Barker. When Barker arrived in Italy he had shot down nine hostile aeroplanes: one as an observer, eight as a pilot. Of his total of fifty-three victories, Barker scored twenty-seven in Italy. His was an exuberant nature and he was a gifted pilot: his instructor sent him solo after one hour's dual. So late in the war, with Smith Barry's influence on instruction, this was phenomenal. During one spell in England he had aerobatted over Piccadilly Circus and escaped retribution by disappearing back to France the next day. He frequently beat up any airfield where he was stationed and delighted in low flying.

Very little has been written about the small and virtually independent British air force that fought alongside the Italians. Among these men were many of colourful personality who generated a special spirit of effervescent enthusiasm that gave the Italian campaign a unique character. Their small numbers nurtured an intimate family spirit that could not have existed at the Western Front at this stage of the war. Also, many American pilots, after training in Italy, joined Italian squadrons to fly Caproni and Ansaldo bombers.

The Italian archives proudly draw attention to the fact that their Service and the British were the only two in the world that planned and followed a line of development as independent arms from those early days. In all other countries, the air forces were mere appendages of the Army and Navy – as they began by being in Britain and Italy – and remained so long after the war: in many instances, until the Second World War.

The squadrons posted to Italy all came by train from France. The first to arrive were 28, equipped with Camels, on which Captain Barker was a flight commander, and 34, which flew RE8s. Barker scored the first British victory in Italy, when, leading an offensive patrol of four, at 10,000 feet on 29th November, five Albatroses attacked. In the next twenty minutes of fighting, seven more Albatroses joined in. He dived on one, which pulled out at 5000 feet, whereupon he fired eighty rounds into it. Its top wing folded back, then the lower one broke off. Between then and the end of the year, No. 28 Squadron shot down eight hostiles and two balloons. No. 66 downed four aircraft and No. 45, two. All three flew Camels.

On Christmas Day the clouds were low, visibility was bad and no operational sorties were ordered. Barker, restless and always ready for any prank, preferably a practical joke, took Lt Hudson with him to the enemy aerodrome at Motta, where he dropped a large sheet of cardboard bearing the message: "To the Austrian Flying Corps from the English RFC. Wishing you a Merry Xmas." They then strafed the hangars and any Austrians who were incautious enough to venture out of doors to see what was going on.

On Boxing Day, soon after breakfast, about thirty enemy bombers, in ragged formation, appeared over and near the RFC airfield at Istrana; thinking it was the home of 28 Squadron. In fact, 42 (RE8) Squadron had left there nine days earlier to make room for 45 (Camels), which had not yet arrived, and only 34 (RE8s) was there. The officers lived in a nearby villa, from where they watched the attack. It was not a good one. Splinters hit all the RE8s but did no serious damage. Less than half the raiding force bombed the target. The rest dropped their loads innocuously about the countryside. One bomb did hit a hangar in which there were two Italian aircraft, which caught fire. The fire set off their ammunition. Braving both the flames and wildly darting bullets, Italian officers and troops wheeled the other aeroplanes out to safety.

An Italian fighter squadron, flying Hanriots (a French marque), was based at Istrana and some of these took off in pursuit of the raiders. Camels of 28 Squadron also intercepted some of these and brought one down.

At 12.30 p.m. a smaller raid struck Istrana. One flight attacked four Gotha bombers escorted by five Albatroses and shot down one of the former. From the accounts of captured pilots, it seemed that these raids were in reprisal for Barker's and Hudson's insult to the Austrian Flying Corps.

The daily task of the RE8s was to find and photograph enemy batteries; and, each aircraft escorted by six Camels, to make long reconnaissances to report enemy movements. Distant reconnaissance was more arduous and risky than an escorted bombing mission, because the RE8s had to fly a devious course and reconnoitre several objectives. The longer flight time, of course, increased the likelihood of interception.

The year ended with the RFC detachment well settled into the routine of their duties, despite the short time they had been in Italy, and the fighter squadrons among them with an appetite already whetted by several victories.

In the last two months of 1917, an Italian pilot, Giovanni Ancillotto, who shot down three enemy balloons in the course of a few days, was so impetuous in one attack that he flew over the balloon, which he had already set alight, and returned to base with his own machine badly damaged and the fabric burned. The official account declares that "This showed the purest spirit of heroism, well deserving of inclusion in the ranks of those selected for decoration with the Gold Medal." Most operationally experienced British and German pilots would have described Ancillotto as careless.

On 8th December a genuinely heroic and accomplished fighter pilot scored three victories: Baracca had brought his total to thirty in seventy-three combats, which earned him the Gold Medal. Another Gold Medal went to Lieutenant Colonel Pier Ruggero Piccio, "of unflagging, cold, astute and serene bravery", who attained his seventeenth victory on 30th November.

Of the attack on Istrana airfield, the Italian account claims that: "This fine victory raised even further the aggressive spirit of our fighters and reduced the enemy's daylight activities to a minimum. The hard lesson, enemy prisoners confessed, deeply hurt their morale. With the 26th December the enemy's short period of air supremacy attained since 1st October definitely ended."

From 28th to 31st December, profiting from favourable weather conditions, the enemy resumed their nightly bombing missions. They dropped 18,500 kilogrammes of bombs on airfields and railway stations, causing large fires, other extensive damage, and many deaths and injuries among servicemen and the civilian population.

It would be fair to end this account of events on the Italian Front in 1917 with a story of Italian gallantry. Sotto Tenente Gino Lisa, a young bomber pilot, had been in many actions against fighters, had often had his aeroplane riddled by bullets, and twice returned from a sortie with it awash with blood. On 15th November he had completed an operation for which he had volunteered, when he saw another Caproni under fighter attack. He instantly went to its aid. In his turn, he was attacked by four fighters. After a long engagement, pitting his heavy and sluggish aircraft against the highly manoeuvrable Albatroses, his machinegunner was hurled out of the aircraft by violent evasive action. The enemy fighters shot the Caproni down and Gino Lisa was killed when it fell to the rocky ground. His Gold Medal was well deserved.

While the air forces were pitched against each other, plans for their future employment were being made, always keeping well in advance of the current situation. The two predominant concerns of the most senior RFC Staff officers were strategic bombing and the combination of the RNAS and RFC into one air Service with its own Air Ministry.

On 11th October 1917, Lieutenant General Sir David Henderson wrote to the Chief of the Imperial General Staff as follows.

"The policy of bombing military objectives in Germany now having been approved, it is necessary to look forward to the operations in this respect next year. It is expected that a very considerable force of aeroplanes will be available, and it is probable that a large part of these will have to operate from centres far removed from the Headquarters of the British Expeditionary Force in France. Preliminary arrangements for the accommodation and supply of the squadrons are already being made by the Commander-in-Chief through the General Officer Commanding RFC, but the full arrangements required for next year will entail a great deal of administrative work and probably a considerable amount of communication with the French authorities. I think the amount of work will be so great that it would probably interfere with the normal duties of the General Officer Commanding RFC in France if the whole burden were thrown on him.

"I consider that a special officer should be detailed to undertake the work of preparation. Whether or not the project for a combined Air Service should come into being, I consider that a force operating at a distance from HQ and detailed for a particular operation of war would carry on more efficiently if under a responsible Commander working directly under the C-in-C, so that the GOCRFC with the Expeditionary Force could concentrate on the particular duties of his Command, which are sufficiently serious to absorb his whole energies."

The effect of this memorandum will be seen later.

Trenchard and others were opposed to a combined Air Service under a separate Air Ministry. Henderson was in favour of it. In retrospect, Trenchard admitted that Henderson had "twice the understanding and insight" that he had. The argument in favour of combining the RNAS and RFC and setting up an Air Ministry are well put in a document among the few surviving Henderson papers, dated 25th September 1917.

"The air services of all belligerents are growing so quickly that no Naval or Military Commander ever has sufficient aeroplanes or pilots. The more they get, the more they ask for and, by the time they have had their demands fulfilled, the enemy has produced more machines, and so they again make fresh requests, and so on.

"Consequently, the Officers Commanding the Grand Fleet, the Eastern Front, and other more detached places, cannot contemplate any air operations outside their immediate necessities until their own wants are satisfied. Consequently a great air offensive against the enemy's supplies, depots and communications will never be proposed or worked out unless it is done by an Air Staff, distinct from the two great Services, which has the power to lay the matter before the War Cabinet, and afterwards plan the offensive and provide the means.

"There are indications that the enemy is at last waking up to the power of the Air and is constructing 4000 large bombing machines. It is vital that we should be before him and have our Bombing Squadrons destroying his aerodromes and machines, as well as his factories, before he annihilates ours. (Night operations are most essential for the destruction of machines, for they will be found in their hangars during the dark hours.) This is the truest method of defence in air warfare, as in all other kinds, and the only effective way of preventing attacks on this country.

"It is essential that a powerful bombing force should be prepared at once, so as to be in the field before that of the enemy. The carrying out of this scheme will necessitate some reduction in other military supplies, but the importance of the offensive to be taken justifies this when the present position of the war is taken into consideration. The Grand Fleet and High Seas Fleet are in a state of stalemate. The armies are in almost an immobile condition. We can only get at the enemy's supplies and communications by air. If we are to do this thoroughly we may cripple him and force him to give place to our armies. If we attack his submarines with aircraft in sufficient numbers we shall relieve the drain on our own resources. If we pulverise his factories and aerodromes we shall stop his raids on this country.

"Germany know this, and information shows that the enemy:—

(a) is reducing output of ammunition

(b) is reducing use of motor transport

(c) has stopped the building of large ships

in order to put all energy into building of submarines and aircraft.

"All this in order to cut our communications and supplies, and to destroy aerodromes and machines and factories before we destroy theirs. It is a sort of 'house that Jack built'. We must destroy the factories that produce the submarines and aircraft, that cut our supplies and bomb London. The enemy wishes to destroy our aerodromes and factories that supply the machines which destroy their factories and aerodromes and so prevent them cutting off our supplies and bombing London.

"The only cure is to put our hangars and shops at the aerodromes underground and bomb the enemy factories out of existence. When this is done the superiority of our forces on sea, land and in the air comes automatically.

"There is no time to be lost. It is a race between Germany and ourselves as to who begins first."

CHAPTER 18

------◆------

1918. Fury in the Firmament

The entry of the United States into the war had thrown the German press into a fervour of expected imminent subjugation by armed multitudes and, from the air, by immense air fleets. The Government and High Command, better informed, knew that many months must pass before America could assemble and train an effective army and air force. Spring 1918, when it was logical to expect the next Allied offensive, was the time for which Germany was preparing. The General Officer Commanding the Luftstreitkräfte knew that, with America's great resources in manpower, materials and wealth, when she did go fully into battle, it would be in mighty strength. He accordingly, in mid-1917, drew up a plan to prepare the Air Service for the coming onslaught. This he called Das Amerika Programm.

On the British side, Henderson expressed his realistic appreciation of aerial warfare and his meticulous accuracy in a paper to the CIGS dated 10th January 1918. He comments, on the publicity automatically given to French pilots when they qualified as "Aces".

"1. 'Mastery of the Air' is not a suitable expression to describe any degree of air superiority. No superiority can be sufficient to ensure that hostile aircraft will not break through on a limited offensive. It is hoped that, with the co-operation of America, our present superiority may be maintained and considerably increased.

"2. The British intend to continue the policy of an air offensive against military objectives in Germany. The name that is applied to these operations does not matter.

"3. The two policies are not quite accurately stated. The French

[236]

announcements refer to the most successful fighting (i.e. fighter) pilots. 'The best flying men' may be employed on more important operations than fighting (i.e. flying fighters). The British system is not to conceal the names and records of flying men, but to publish them after they have been examined and approved by the Commander-in-Chief and the King. My own opinion is that the British policy is the better, both as a matter of justice and as tending to keep up the level of efficiency throughout the Force whatever work the pilot may be engaged on.

"4. No amount of superiority, in number or in quality of machines and pilots, can hope to secure absolute mastery of the air, nor can even a dominant position be assured unless the organisation of the Force and the spirit and skill of the pilots keep pace with the provision of mechanical appliances."

The sentiment about justice in his third paragraph is admirable, but he overlooks the equal merit and claim to recognition of observers and aerial gunners.

There are two points to make about Henderson and Trenchard in which the attitudes they took differed from those customarily attributed to them. The DH4's successor, the DH9, also a two-seater bomber, was intended for long-range bombing. Geoffrey de Havilland was uneasy about this: its performance was not adequate and would put the crews' lives in jeopardy. He said it should not be put to this task.

Henderson demurred. "It may be inadvisable to attempt any bombing until a better machine is produced, by which time again the German machines may also have improved. If the DH9 is fit for day bombing, then the more of them we use the better."

Trenchard totally disagreed and put his point with witty sarcasm. "I want to bomb Germany, but please remember that if we lose half our machines doing so, the good morale effect which is three-quarters of the work will be on the German side and not ours."

Trenchard had been knighted in the New Year Honours List. Henderson was no longer at the War Office. He had moved to the newly formed Air Ministry as Deputy Chairman of the Air Council, and 1st April 1918 had been set as the date for the amalgamation of the RNAS and RFC into the Royal Air Force.

Political pressure of the shabbiest sort had been exerted on Trenchard to induce him to return to England and become Chief of Air Staff. Like Henderson, three years earlier, he put the good of the Service before his own advancement and would have preferred to remain in France. Insistence by Lord Northcliffe and his brother, Lord Rothermere, both newspaper-owners with huge political power, proved too high and he had to accept the promotion. As Henderson had done, he left the field for an office desk. His place was taken by Sir John Salmond.

By March, Trenchard, sickened by further political intrigue, resigned. He was astonished when the man chosen to replace him proved to be Sykes. He was offered the choice of three new appointments, but elected a fourth: command of an enlarged bomber force to be stationed in France.

None of these basic changes made any noticeable practical difference to the lives of the aircrews at the Western or Italian Fronts.

The annual winter reduction in air activity ended with the great German offensive that began on 21st March. By then the Luftstreitkräfte had grown to 153 Fliegerabteilungen, 80 Jastas, 38 Schlastas, and 24 Bombengeschwader that comprised 24 Staffeln.

Of these, pitted against the RFC were 49 Fliegerabteilungen, 35 Jastas, 27 Schlastas and 4 Bombengeschwader: 730 aircraft, of which 326 were single-seater fighters.

Ranged against these the RFC had 579 aircraft serviceable, 261 of them single-seater fighters. The fighters were mostly SE5as and Camels. There were five Bristol Fighter squadrons and one flying the Sopwith Dolphin.

The RFC's air activity preliminary to the offensive consisted of bombing enemy airfields, and targets that would impede troop movements: railways, bridges, roads. Many fighters were sent on low-level bombing raids.* The RFC C-in-C, as always, took it for granted that aggression would overcome the defence. The Luftstreitkräfte, also as usual, relied on well-planned defensive tactics. The British casualties, although the Staff appeared not to recognise the fact, proved heavier than the enemy's.

March 18th saw a battle between five DH4s escorted by twenty-four Camels and SE5As, against thirty Albatroses, Pfalzes and Fokker Triplanes led by Richtfhofen, plus eight Staffeln flying the same variety of fighters. Badly outnumbered, the British did well to lose no more than two DH4s, two SE5As and five Camels. But it was a heavy defeat nevertheless: the Germans lost one Albatros.

The offensive began in poor weather, with fog that hampered flying. The Germans' initial advance drove many squadrons to retreat on the second and third days to aerodromes further behind the line. Naval 5's base was shelled and some of the squadrons, including Sholto Douglas's 84, moved with only an hour or two to spare before the enemy overran them.

---

* As Messerschmitt 109s and Hurricanes were used as fighter-bombers from late 1940.

Eighty-four Squadron had been strafing day after day. It was a rich time for Beauchamp-Proctor. He had shot down his first enemy aircraft, a reconnaissance type, on 3rd January. He got his first fighter, a Fokker Triplane, on 17th February and soon shot down four more fighters and was given an MC. On 1st April 1918, the day on which the RAF came into being, he was promoted to captain and given command of a flight.

Sholto Douglas insisted on his squadron operating in three flights of five aeroplanes and staying in formation. The individualists objected to this, and he issued an order that no one was to break formation to snatch an easy chance to shoot down an enemy. The initiative for any attack must come from the leader. As this would be the most experienced pilot, he had the best prospect of shooting down the chosen victim. McCudden followed the same system. This style of fighting aroused resentment and jealousy. Some of the other pilots objected that the leader scored most of the kills while they guarded him. The leaders pointed out that most new pilots were shot down when a formation broke up. The way to give them experience was for all to hold formation, with the raw men guarded by the experienced, who also protected the leader.

McCudden had ended 1917 with two dazzling displays of marksmanship and the style of flying that combat demanded. On 23rd December he shot down four hostiles, and on the 28th he shot down three LVGs in twenty minutes. The first of these he took by surprise from astern and below; and afterwards wrote: "I hate to shoot a Hun down without his seeing me, for although this method is in accordance with my doctrine, it is against what little sporting instincts I have left." The second one, he destroyed at a range of 400 yards. He said he knew this estimate would be disputed, but was adamant about his accuracy in judging range. Most pilots, he said, underestimated the range at which they opened fire, and novices usually fired their first bursts from twice the distance they thought they were from their target. This was often said in the 1939–45 war, with equal truth, and was a frequent topic of squadron and flight commanders' observations on newcomers in action.

Being an ex-mechanic, McCudden was very conscious of the disparity between the performance of aircraft of similar type. He had lately been irritated to see Rumplers passing him when above 15,000 feet, and found out that their original engine had been replaced by a more powerful one. A new engine was being put into the SE5A, and he obtained a set of high-compression pistons for it. "I was very keen to see the Rumpler pilots' hair stand on end as I climbed past them like a helicopter." What is as interesting as his attention to mechanical detail is the fact that he knew the word "helicopter". None had yet been built and he did not live to see their advent.

He had the strict and intolerant views about some matters that probably arose from his training as a regular ranker. On 24th January he shot down an artillery-spotting DFW at 12,000 feet. "This crew deserved to die, because they had no notion whatever of how to defend themselves. Which showed that during their training they had been slack and lazy. They probably liked going to Berlin too often instead of sticking to their training and learning as much as they could. I had no sympathy for those fellows." This was his forty-third victory.

But he could also be admiring. On patrol, he saw a DFW two-seater and led his flight onto it. It was below cloud, at 4000 feet, so he detached his companions to wait above cloud in case it escaped him and went down to engage it. His Vickers was out of action, so he fired his Lewis. For five minutes he fought the enemy down to 500 feet. "At last I broke off the combat, for the Hun was too good for me and had shot me about a lot. Had I persisted he certainly would have got me, for there was not a trick he didn't know, and so I gave that liver-coloured DFW best." There was an Albatros with a green tail whose pilot he had often seen in action, displaying great skill and shooting down British aircraft, including some of McCudden's squadron. One day his flight met a patrol led by Green-tail and McCudden put a burst into him. The Albatros burst into flames. The German pilot had tumbled out "and was hurtling to destruction faster than his machine. I now flew on to the next Albatros and shot him down at once. I must say the pilot of the green-tailed Albatros must have been a very fine fellow. I had many times had cause to admire his fighting qualities. I only hope it was my first bullet which killed him."

On 16th February he again shot down four. At the beginning of March he was posted home, where he was promoted to major. In April his Victoria Cross was gazetted.

On 8th July he was appointed to command 60 Squadron, one of the best. The following day, when he took off to join it, his engine stopped; and, going against standard procedure, he turned back. The aeroplane stalled into the ground and he was killed. He had scored fifty-seven victories.

Mannock, by this time, with an MC and bar, had spent some months in England instructing, and flying FE2s on wireless testing. On a chance meeting in London with Henderson he had been outspoken about his boredom and frustration. In consequence, he was posted early in March to 74 Squadron, which was working up for the Front, as a flight commander. He set about making himself noticed by organising sing-songs in the officers' mess, leading at the top of his voice and "playing" with drumsticks on a collection of cans, tankards, pots, pans and glasses tied to the back of a chair.

The squadron arrived in France on 1st April 1918. His approach to fighting was highly analytical and – in keeping with his boring conversational style – he used to subject his pilots to a thorough and helpful analysis of every engagement. He was an inveterate do-gooder and busybody, but kind with it. One of his many concerns about his inexperienced pilots was with the effect on them of seeing aircraft shot down. In those days of easy conflagration and no parachutes, this could rapidly unnerve a new man to the point where he was soon a candidate for admission to one of Henderson's hospitals for psychiatric treatment; and that was crude enough in itself to excite all sorts of new traumas. Mannock's own mental state was none too well balanced. He soon became obsessed by the sight and stench of burning aeroplanes and of the men in them. One accident gave his growing insanity a shove closer to the edge of disintegration, and his nose a close-up, when a comrade crashed on the aerodrome and was incinerated.

One evening, after Cairns, a particular friend, had been killed, he was in great distress and mental turmoil. He announced that the mess must give Cairns "a good send-off". After the usual rowdy games he rose to make a speech. "To Captain Cairns and the last dead Hun. Sod the Huns." It was probably the best such oration of the war: brief, full of feeling and apposite. Next morning he was jovial once more: classically manic-depressive.

On 18th June, while on leave, he was awarded a second bar to his DSO and made a major. Posted to command 85 Squadron, he went first to say goodbye to 74. In the mess that evening he broke down and cried. It was his frank revelation of his feelings that endeared him as much as it often annoyed.

He found morale on 85 low. His predecessor, Billy Bishop, was a loner instead of a leader. He interviewed each pilot and got rid of those whom he found suspect. Three were Americans, all of whom he kept: Elliott White Springs, Larry Callaghan and John Grider. He began at once to teach his squadron to fight as a team and his pilots were immediately inspired with eagerness. The first time he took them out to fight, he selected three to go with him as decoys. Two other flights followed at different altitudes above. At 8.20 p.m. Mannock sighted ten Fokker DVIIs approaching. He dived with his other decoys and the enemy followed. When he signalled, five of the flight next above also went for the Fokkers. Then the third flight came down and took them by surprise. The fight began at 16,000 feet and ended at 2000 feet.

Mannock's squadron had no losses. The Germans lost five aircraft, of which Mannock shot down two.

When he heard of McCudden's death he became more neurotic, depressed and full of forebodings about his own end. To assuage his fears he indulged in a week of solo sorties. He attacked every enemy he saw and could reach. After a week of hysterical destructiveness he seemed calmer. He now became obsessive about the neatness of his turnout. He talked openly about having a strong premonition of death: hardly the way to encourage his subordinates.

He invited a friend to lunch soon after equalling Bishop's score of seventy-two. The friend said there would be a red-carpet reception for him after the war. Mannock did not think so: "There won't be any after the war for me." Later, when his guest mentioned a flamer he had shot down, Mannock asked: "Did you hear the swine screaming? That's the way they'll get you if you're not careful." The RAF has a tradition of black humour, but not of morbidity. "When it comes, don't forget to blow your brains out." The reluctant laughter at this gruesome "joke" soon stopped when Mannock described a burning aeroplane: many pilots carried a revolver to commit suicide if they were going down on fire.

He was shot down soon after, with the top RFC/RAF score of seventy-three; and given a posthumous VC.

The Escadrille Lafayette was disbanded on 18th February 1918. It took on a new identity, l'Escadrille Jeanne d'Arc, and the American pilots, who were transferred to their own country's flying Service, were replaced by Frenchmen.

Raoul Lufbery, transformed into a major in the US Air Service, was sent to the newly arrived nucleus of the 94th Pursuit Squadron. It was commanded by another ex-Lafayette, Major John Huffer, but "Luf", America's most famous military pilot, at once became its pivotal member. The French were tardy in supplying the aircraft that America had bought. Presently some old Nieuports were delivered: but there were no machineguns for them.

Eddie "Rick" Rickenbacker had arrived in France in 1917 as driver to General Pershing, the American Commander-in-Chief. He applied to join the Air Service, but the general was reluctant to let the famous racing driver go. Rickenbacker had a chance encounter with Colonel Billy Mitchell, the Commander of the Air Service, on whom he made such a good impression that Mitchell persuaded Pershing to release him. In January 1918 he was a newly commissioned lieutenant at Issoudun, learning to fly. Race driving had made him an excellent judge of speed and distance. He proved a natural flyer and soon became an accurate shot. In March he was also posted to the 94th. Among the pilots were Reed Chambers and Douglas Campbell, two names to remember.

On 6th March, two days after joining the squadron, Rickenbacker accompanied Campbell, under Lufbery's leadership, on the 94th's first flight over the lines. They took off at 8.15 a.m. and climbed to 15,000 feet. Anti-aircraft shells were soon bursting near them and pitching their aeroplanes about. On landing, Lufbery asked them what they had seen. They had seen no aircraft, and were astonished when he told them that a formation of five Spads had crossed under them before they reached the lines; another five Spads had passed within 500 yards a little later; four Albatroses were two miles ahead when they turned for home, and an enemy two-seater closer to them at a height of 5000 feet. This was their first, and chastening, lesson in how difficult it was to discern other aeroplanes until one had learned how to fly and keep a sharp all-round lookout at the same time. It was as well they had not had a brush with the enemy: they were still without guns.

There was another American squadron at the Front: the 95th, under Captain James Miller. But the 94th were sent on a course at the gunnery school and by 10th March their aeroplanes had been armed; while the 95th took their turn at gunnery school.

On 13th April Captain Peterson, Lieutenant Rickenbacker and Lieutenant Chambers were ordered to make the American Air Service's first operational patrol of the war, at 16,000 feet between Pont-à-Musson and St Mihiel. They were woken at 5 a.m. and found a heavy mist. Peterson sent the other two up on a weather reconnaissance. After orbiting the field twice at 1500 feet they saw Peterson coming to join them and they all climbed to 16,000 feet. Before they had reached the start of their patrol line, Rickenbacker and Chambers saw Peterson glide down, assumed that he had engine trouble – they had arranged no signals – and carried on. After going up and down the patrol line twice they turned for home. The area was under fog and they had thirty minutes' fuel. They now understood why Peterson had turned back. They dived through the cloud and fog, and lost each other. Rickenbacker had to go down to 100 feet to find clear visibility and scraped home with his tank nearly dry. So did Chambers.

Campbell and Alan Winslow had been detailed to stand by from 8 a.m. until 10 a.m. Soon after Rickenbacker landed a report was received that there were two enemy aircraft in the vicinity. Campbell and Winslow took off. Winslow shot one down in flames and Douglas Campbell forced the other down. Neither enemy pilot was badly hurt. The US Air Services' first two victories brought the successful pilots telegrams from home and telephone calls from England and several Allied flying formations in France.

There were six members of the squadron who had been trained by foreign air forces and had already shot down five or more aircraft each; Lufbery, Baynes and Putnam, flying with the Escadrille Lafayette, Warman, Libby and Magoun with the RFC. But Campbell had done all his service with the USAS and been trained by it.

Later in the month Rickenbacker and James Norman Hall, a former Lafayette, intercepted a Pfalz and dived on it from 1000 feet above. Rickenbacker moved to one side to cut the Pfalz off, while Hall attacked. The surprised Pfalz turned for home, whereupon Rickenbacker fired and killed the pilot. On their way back to base they met flak. Rickenbacker would have turned away, but Hall flew on, so he followed. They then stunted for ten minutes among the shrapnel, in the euphoria of their victory. Rickenbacker's delighted comment on the congratulations of their comrades was "No closer fraternity exists than that of air fighters in this great war." It was extraordinary that a pilot with only four months' flying experience should shoot down a hostile aircraft so late in the war, when they were a great deal more difficult to hit than the BE2 or DFW B1 had been, or the Voisin or Taube. The French wanted to award both Americans the Croix de Guerre with palms, but American officers were not at that date allowed to accept foreign decorations.

Rickenbacker said that the 94th was most anxious to prove its worth, because everyone suspected that the British and French, with three years' fighting behind them, looked on American pilots with amusement and polite contempt. This was sensitivity: few experienced pilots would adopt that attitude. Everyone recalled his own novice days.

On 2nd May, the 94th suffered their first loss. Captain Peterson was on patrol with three others, among whom was Chapman, when they fought five Pfalz. One of these shot Chapman down in flames.

Lieutenant Elliott White Springs was one of several US Air Service pilots who were sent to serve with the RFC and, as it had become when he joined in April 1918, the RAF. He was on 85 Squadron, under Billy Bishop, and wrote enthusiastically in his diary about his welcome to England. "It is surprising how well Americans are getting on over here — much better than I expected. Everyone that has come in contact with the British swears by them. And the British will do more for you than they will for their own troops. Every club in England that is open to English soldiers is open to Americans. If the English soldiers are entitled to special prices, so are we. We ride on their trains at half fare and we are entitled to anything we want from their canteens. Yet when the

American commissary was opened up in London, the first rule they made was that no one could buy anything who wasn't in American uniform. As for me, I'm for the British and I don't care who happens to know it."

On 19th May he was one of the three Americans who went to France with Bishop, flying the SE5A in a formation of nineteen. These long flights were still fraught with uncertainty. One pilot dropped out with engine trouble after fifteen minutes. He took off again, crashed, and started once more with a new aircraft next day. Another crashed soon after crossing the Channel. A third made a heavy landing and broke his aeroplane. The squadron had, in addition to the Americans, two New Zealanders, two Australians, six Canadians, two Scots, one Irishman and six Englishmen.

On 4th June Springs landed on an aerodrome occupied by an American squadron near Dunkirk and said of it: "These boys are certainly down in the mouth. They think the Hun has won the war and are worrying about their baggage and girls in Paris. I asked some of them to dinner, but they haven't much transport and have to account for every drop of petrol. I offered to send a car for them, but they didn't show much enthusiasm. I gather that Uncle Sam is pretty stingy. Me for the RFC. There's one thing about the British that I like – they realise the importance of morale. The British try to build it up, the Americans try to tear it down." What he meant was that the RFC had learned the value of loosening discipline, whereas the US Air Service was still operationally inexperienced and on some units discipline was too strict. He would not have made the same derogatory comment about the 94th or 95th Squadrons, new though they were to campaigning. With men like Lufbery and other former Lafayette and RFC pilots among them, and someone with Rickenbacker's character, the atmosphere was as happy as on the average RFC or RAF squadron.

On 23rd June Sir John Salmond came to tea with the squadron. Bishop had been posted home and Salmond wanted McCudden to fill his place. The squadron opposed this. McCudden had the reputation of shooting down all enemy aircraft himself and giving nobody else a chance. They asked for Mannock, "said to be the best patrol leader at the Front. He plans the day before and rehearses on the ground. He plans every manoeuvre like a chess player and every man knows his job. He has marvellous eyesight but only one eye." The truth was that Mannock did not see very well with his good eye. Springs had something else to say, which fits Whitehouse's complaints. Some of the officers did not want Mannock "because he is an ex-Other Rank and his father was a sergeant major. These English have great ideas of caste." Maybe: but

McCudden was an ex-ranker and nobody seems to have given that as a reason why he should not have the squadron.

On 7th July, Captain Baker, acting CO until Mannock arrived, put Springs up for a DFC, but this was turned down on the grounds that four victories were not enough.

By the time August was well advanced Springs was entering in his diary: "I don't know which will get me first, a bullet or nervous strain." Ground strafing was 85's main task just now and the pilots, flying in pairs, were each making three or four sorties a day. On 27th August: "We've lost a lot of good men. It's only a question of time until we all get it. I'm all shot to pieces. I only hope I can stick it. I don't want to quit. My nerves are all gone and I can't stop. I've lived beyond my time already. Here I am 24 years old, I look 40 and feel 90."

The erosion of mental and physical strength, energy and resilience that afflicted pilots during arduous periods at the Front, and most oppressively at this period if they were heavily committed to strafing, was not, for most of them, caused by fear of death. It was the daily flinching from it that wore out the nerves. Its effects grew on them day by day, ravaged the constitution and undermined their sanity. Most of them were too young ever to have taken any aspect of life seriously, but they found the demands of their task were causing so radical a change of attitude that many wondered if they would ever be carefree and joyous again. The end of the war would come too late. They would be changed men and remain permanently different from the men they had been.

Almost the last entry in Springs's diary is, in this context, as poignant as any lines of wartime poetry written by any soldier or airman who had served long at the Front. "Sooner or later I'll be forced to fight against odds that are too long ..." Or perhaps a stray bullet from the ground would kill him, or his engine would cut while he was low over the trenches on a strafe. "*Oh, for a parachute!* The Huns are using them now."

At about 10 a.m. on 19th May, an enemy photographic reconnaissance aircraft was reported approaching Toul. The only pilot ready to take off at once was Lieutenant Gude, who was ordered to do so immediately. The French anti-aircraft batteries had been firing, but ceased. It was assumed that they had hit the Albatros two-seater, which was descending erratically. When it was down to about 200 feet, however, it straightened out and turned for the lines. Gude attacked it at once; his first combat. He was too far away to hit the enemy, used all his ammunition and returned to base. The anti-aircraft fire resumed, but ineffectually.

Lufbery had been watching and now dashed to the hangars on his motorcycle. His own aircraft was being serviced, so he took another. In five minutes he had overhauled the enemy machine, at 2000 feet and well within view from the airfield. The onlookers saw him dive, fire several short bursts and follow the enemy down to 200 feet. He swerved away with his guns apparently jammed. He circled the aerodrome, evidently cleared his guns, and attacked the enemy from astern.

Lufbery's Nieuport suddenly erupted into flames. A few seconds later he was seen to jump out of it rather than be roasted alive. Examination of the wreckage showed that a tracer bullet had hit the petrol tank. Rickenbacker recalled that only a few days before he had asked Lufbery what he would do if his aeroplane was set alight: jump or stay. Lufbery had replied that he would stay, because there was always the chance of sideslipping down to the ground and thus fanning the flames away from the cockpit.

A pilot of a French squadron based nearby, who had seen what happened, took off at once to attack the Albatros. The German machine's first burst shot him through the heart. A second French fighter did shoot it down, it crashed on the Allied side of the lines and its crew was taken prisoner.

Douglas Campbell had taken off to avenge Lufbery and came back after an hour to report that he had fought a Rumpler, killed the observer, wounded the pilot and brought it down, also behind the Allied lines. Hundreds of mourners from all branches of the Allied Services attended Lufbery's funeral the next day.

On 6th May the 95th Squadron had finished its gunnery course and rejoined the 94th at Toul, eighteen miles from the Front and near Nancy. The accommodation was good: stone buildings and steel hangars.

Hartney, in the meanwhile, commanding the 27th Squadron, had returned from Canada and America, spent six days in England, and arrived in France on 18th March. On 1st June, the 27th and 147th Squadrons also arrived at Toul with their Nieuport 28s.

The 94th's Commanding Officer gave them a most hospitable welcome. Major John F. M. Huffer, born in France to American parents, had never been to the United States but spoke English fluently. Along with the courtly manners he had learned in France he had the inherent American generosity. He invited all three of the other squadrons, and all twenty-four of the American nurses in the neighbourhood who were off duty, to a party, and sent a young officer to Nancy to fetch as many respectable young French ladies as he could persuade to venture among the strange foreigners, with chaperones who included the mayor's wife. Hartney described the party as a combination of American barbecue and gala French fête. Dancing went on until 4 a.m. The festivities were concluding with everybody gathered on the aerodrome singing

"Lafayette, We Are Here", when enemy bombers gatecrashed. Amid falling bombs there was a rush for the air raid shelters. Nobody was hurt. It was as pleasant a baptism of fire for the 27th and 147th as anyone could wish for.

At 6 o'clock that very morning the two maiden squadrons began patrolling. The first off the ground consisted of the 27th's Lieutenants Grant, McElvain, Hunt and Raymond, led by Lieutenant Taylor of the 95th. The rest of the day's patrols were also led by pilots of the 95th, while the 94th introduced the 147th to the front line: which was eminently sensible and an excellent way of promoting good relationships between all four squadrons.

Two days later an enemy aircraft dropped a photograph of the airfield with the message "Welcome 27th and 147th. Prepare to meet thy doom."

The Americans were only too keen to meet the enemy.

Among others of their countrymen who were gaining experience on detachment to British squadrons, was a whole flight of American pilots and mechanics, some fifty in all, operating with Sholto Douglas's 84 Squadron at Bertangles. All three flight commanders, as it happened, were South Africans, one of them being Beauchamp-Proctor. Douglas was at first reluctant to have his squadron disrupted by the arrival of so many from a different country's air Service, but soon enjoyed their presence. He described them as cheerful and courageous and as having a good effect on morale. They also joined in eagerly at rugger, which particularly endeared them to their temporary CO. Among the pilots was George Vaughn, who ended the war as one of the most successful pilots in the USAS, with thirteen kills. "A thoroughly pleasant man and a first-rate shot." He scored six kills and shot down several balloons during his time with 84 Squadron. When the first batch of fifty left after three months, another lot took their place. They were all a fine lot of men, said Douglas, and he disliked having to part with them.

CHAPTER 19

———◆———

1918. Victory

In Italy the new atmosphere created in the Army consolidated on the Piave, uplifted by British support and patriotic propaganda, affected also the Air Corps, spurring it on to greater efforts. In the first four months of 1918 the bomber squadrons systematically attacked enemy airfields, ammunition dumps, barracks and railway lines by day and night. The enemy's air offensive during the first two months, while German bomber squadrons were still there, was fierce.

The Air Corps's contribution to preparations for the Battle of the Piave River was not limited to conventional operations. On the night of 30th May 1918 a Voisin with a silenced engine deposited Lieutenant Camillo De Carlo, an air observer, behind enemy lines, where he remained for three months. His espionage work in determining enemy plans and where the Austrian attack was to be made earned him the Gold Medal. He was able to discover, a few days in advance, the date when the attack was to be launched. By laying out laundry in a meadow as though it were drying in the sun, he sent coded signals which reconnaissance aircraft duly photographed.

At 3 a.m. on the night of 15th June a heavy Austrian artillery barrage announced the beginning of the battle. On this Front the Italian Air Corps had 221 fighters, 56 bombers and 276 reconnaissance machines. The French had 20 reconnaissance aircraft. The RAF had 54 fighters and 26 reconnaissance types. The infantry assault was supported by low-level Austrian bombing and strafing, to which the Allies responded with fighter patrols in large numbers. On that day 37 aircrew on both sides were killed. Between 15th and 25th June, Italian and British fighters shot down 107 enemy aeroplanes and 7 balloons. On the last day of that period, Lieutenant Flavio Barracchini scored his 25th victory.

On 19th February the Italian 18th Bombardment Group, comprising three Caproni squadrons, had arrived at the Western Front. On their first night, the enemy attacked their airfield at Longive-Ocheley and damaged three aircraft. During March, April and May they flew fourteen operations, in which sixty-five aircraft took part. Their targets were mostly railway stations and airfields. In April they moved to Villeneuve and in August to Cherminus. Between their arrival and the end of the war, they flew sixty-eight operations, lost seven aircrew killed and suffered fifteen wounded.

On the Italian Front, American pilots, flying with Italian bomber squadrons, took part in operations for the first time on 20th June. Some of them won Italian decorations. Two of these, Lieutenants Coleman de Witt Fenafly and James Bahl, were posthumously awarded the Gold Medal. On the afternoon of 27th October 1918, during a bombing raid, as first pilot on a Caproni aircraft, Fenafly was attacked by five enemy fighters. Instead of avoiding the unequal battle by landing, he chose to fight. His gunners shot down two of the enemy and continued firing after their aircraft broke out in flames, until it was destroyed and the entire crew perished. On the same mission, five enemy fighters also attacked Bahl's Caproni. His gunners, too, sent down two enemy aircraft before their own caught fire, then continued shooting until it crashed, killing the crew.

The RAF squadrons in Italy had been hard worked throughout. In March, 42 Squadron had returned to France. Joubert de la Ferté, who commanded the RAF in Italy, found that one squadron of RE8s could not cope with all the bombing and reconnaissance that was needed. Six Bristol Fighters, flown by crews with no operational experience, were sent from England and added to 28 Squadron as Z Flight, to work up to operational standard. At the end of March they were transferred to 34 Squadron.

On 30th March twenty-year-old Lieutenant Alan Jerrard of 66 Squadron won the only VC awarded in Italy. He took off that morning with his flight commander, Captain Carpenter, and Lieutenant Eycott-Martin on an offensive patrol. They met four Albatros DIIIs escorting a Rumpler and attacked at 13,000 feet. Carpenter and Jerrard each shot down one fighter. Presently Carpenter and Eycott-Martin saw, from 6000 feet, that Jerrard had gone down to about 50 feet and was attacking, one after another, six Albatroses that were trying to take off. The combat report states that there were nineteen enemy aircraft milling around. Other details are confused. Both Carpenter and Eycott-Martin were in everal combats and returned without Jerrard, who was shot down and aken prisoner. He was awarded his decoration for his bravery in sustained low-level attacks against much superior numbers.

On 19th June, the Italian 91st Squadron was ordered to make a machinegun attack on the enemy trenches. Baracca set off with two other pilots who were skilled ground strafers, Lieutenants Osnaghi and Costantini. Half an hour later Osnaghi returned to tell Colonel Piccio, he wing commander, that he thought Baracca had been shot down. They had made their attack. He was on Baracca's right and fifty metres above him, when he saw a tongue of flame lick out from Baracca's machine. At the same time it zoomed up and he lost sight of it. Five days later burned-out remains of Baracca's Spad were found. He had scored thirty-four victories.

Ruffo de Calabria succeeded him as squadron commander, survived the war and made his last flight nine days before the Armistice. His score was twenty. Silvio Scaroni also survived, with twenty-six kills to his name; and so did Piccio, with twenty-four.

The operations and combats of the British squadrons were the same as those on the Western Front, with the exception that contact patrols were not necessary. In less than twelve months, the three Camel squadrons brought down 367 enemy aircraft. Only thirty-two Camels were lost on the enemy side of the lines. Nineteen British pilots were killed in action, four were wounded in action but returned safely, nineteen were taken prisoner and survived the war.

The squadrons in Italy comprised the same variety of nationalities as in France. On both Fronts, pilots' average age was strikingly younger than three years earlier. No. 45 Squadron was typical. Major J. A. Crook, MC, the CO, was aged twenty-one. His flight commanders, Captains N. C. Jones, R. J. Dawes and C. E. Howell, were twenty-five, twenty-one and twenty-four respectively. The others were mostly between eighteen and twenty. Five officers were from the United Kingdom, fifteen from Canada, Australia and New Zealand, and three, Second Lieutenant Charles Gray Catto and Lieutenants Jay Rutlidge O'Connell and Max Gibson, from the USA.

Forty-five Squadron was honoured by a visit from His Royal Highness The Prince of Wales that did not go quite as the lieutenant colonel commanding the wing had expected. The squadron borrowed a band from a famous British infantry regiment for the occasion. The musicians found a barrel of vermouth in the kitchen. Their playing became progressively "more unusual" as it was described at the time, until the bandsmen and their instruments were slung into a lorry and driven off. The musicians were evidently not the only excessive imbibers. When

Captain J. Cottle, who had by then taken command of A Flight, and the other members of the dawn patrol went into the mess the following morning, they found the Medical Officer sound asleep, wedged on a lavatory seat.

Shortly before this, Cottle had shot down Oberleutnant Linke-Crawford, the third-highest-scoring Austro-Hungarian ace, who had thirty victories to his name at that time. Leading a patrol with Catto and eighteen-year-old Lieutenant F. S. Bowles, Cottle saw three enemy fighters overhead, which at once attacked. Their leader – Linke-Crawford – came in from astern of Cottle, who made the extraordinary decision to loop. Usually this is a manoeuvre that invites a burst of gunfire, because it gives the attacker a full plan view of the target. Its unorthodoxy apparently surprised Linke-Crawford, who, in his faster-climbing machine, looped over Cottle; who shot his aircraft to bits.

On 15th July Barker had been promoted to major and given command of 139 Bristol Fighter Squadron, which had been formed by adding six Brisfits to the original six in Italy. Barker took his Camel with him and brought down six enemy aircraft with it before being posted back to France; where he shot down several more Germans.

Early in the year a parade had been held at Istrana to present the British Military Cross to d'Annunzio. He continued flying on operations whenever possible and was finally awarded the Gold Medal for Military Valour and promoted to lieutenant colonel.

The Armistice between Italy and Austria was declared on 4th November 1918, a week earlier than the Armistice on the Western Front.

In France, the pace and intensity of the air war was maintained until the end. On 3rd October Degelow shot down a French Spad and on 30th October a British DH4. He was awarded the last Blue Max of the war.

Leo Leonhardn received his on 2nd October and was at the Front until the 15th, before being given command of a training school.

Von Röth, who rivalled Gontermann as an expert in destroying observation balloons, and followed the same techniques of attack, brought down five balloons in flames on 29th May. His total, for which he was decorated with the Blue Max, was twenty balloons destroyed.

Von Greim, who made his name by knocking out tanks, was not given his first opportunity to attack these new inventions until 1918. His first attempts to bomb them and hit them with armour-piercing bullets were frustrated by heavy ground fire. The basic difficulty, however, was to find a tank to attack. There were very few, and they were widely scattered. Camouflage made them difficult to espy. This meant that the whole sortie had to be flown very low: which exposed the aircraft to incessant fire from machineguns and quick-firing cannon. The best method of attack had also to be found empirically. It was not necessary to pierce the tank and kill or wound the crew. To stop it, only the tracks needed to be hit. It was easier to do this by machinegun fire than by trying to judge when to drop a small bomb. He was not trying to shoot men but to bring a mechanised vehicle to a halt. These behemoths amazed him by the speed with which they could turn and avoid his fire. He found that the most effective attack was a vertical dive from 600 metres with both guns firing, and a pull-out that made the aircraft creak as though the wings would be torn off. In this way he was out of sight of the tank gunners until the last moment and had the tank broadside-on at his mercy. It was the spring of 1918 before he perfected this method. In October he was given the Pour le Mérite.

The manner of Manfred von Richthofen's death remained in some doubt until the 1960s, when it was thoroughly researched for the first time. On 21st April 1918, in a dogfight, a Canadian pilot, Wilfred May, found that the guns of his Camel had jammed. He left the fight and headed for his base, Bertangles, at low level. Richthofen chased him. May kept zig-zagging and Richthofen fired at every chance. They crossed the British trenches in an Australian sector. Another Canadian, Roy Brown, was chasing Richthofen, shooting at him. Eyewitnesses confirmed that Brown broke off after firing a last long burst. Machineguns on the ground and rifles were also being fired at Richthofen. Brown thought he saw his tracer hit the rear fuselage of Richthofen's aeroplane. Richthofen crashed and was found dead in the wreckage. Brown was awarded a bar to the Distinguished Service Cross he had won in the RNAS.

Investigation half a century later, and an interview with Brown, has left no doubt that the man who shot Richthofen down was an Australian machinegunner, Robert Buie, of the 53rd Battery, 14th Australian Field Artillery. Buie was manning a Lewis gun on anti-aircraft lookout to protect the big guns, saw the enemy aircraft overhead, shot at, and hit, it and was always convinced that it was he who shot it down.

The previous autumn, Werner Voss had been shot down by a young pilot of 56 Squadron, Arthur Rhys-Davids. Voss was up alone in a Fokker Triplane at dusk when he attacked a British straggler. Six of 56's most experienced pilots, led by McCudden, were above and out of sight. Their SE5As dived faster than the Triplane and caught it up. McCudden and Rhys-Davids separated to cover both flanks, two others stayed above and two astern. Voss turned suddenly to face his enemies and flashed past them at a closing speed of nearly 200 m.p.h. without being hit. But he was boxed in and although he twisted and switchbacked for some minutes, he could not break free. Any of the British pilots might have killed him, but it was Rhys-Davids who got the chance and took it.

He, in turn, disappeared on a dusk patrol not long afterwards, with twenty-three confirmed kills.

No. 55 Squadron, flying DH4s on the Headquarters Wing commanded by Lieutenant Colonel Cyril Newall, had been engaged on strategic bombing since early 1917. In October that year Newall had been given command of a new wing, No. 41. Its first squadrons were No. 55, whose DH4s had been on daylight operations since April, and No. 100, which, since the same month, had been flying FE2bs by night. It was typical of the state of aircraft supply that these latter, obsolescent lattice-tailed fighters, should still be flying against the enemy as makeshifts for a purpose different from that for which they had been designed. Also allotted to the wing was a Naval unit, "A" Squadron, which had the huge 100-foot-wingspan Handley Page o/100.

Late in April 1918 the first two of these became the nucleus of a further new formation. The Independent Force, comprising four day (Nos 55, 99, 104 and 110) and five night (Nos 97, 100, 115, 215 and 216) squadrons, was stationed at several airfields in eastern France, under Trenchard's command, for the purpose of bombing Germany and enemy targets in France and Belgium. The IF began operations under its new identity in June 1918. Not all the squadrons had by then joined it. Nos. 97, 115 and 215 did so in August. Had the war not ended when it did, American, French and Italian squadrons, all to be commanded by Trenchard, would soon have been added to it.

Whether they raided by night or by day, the bombers did so in the teeth of an array of defensive measures that had been greatly augmented in recent months. Fighters – helped after dark by searchlights – patrolled across the lines of approach to the most obvious targets. Flak was sited intelligently and in abundance, along the routes and around the targets. Barrage balloons forced attacking aircraft to fly high and thus reduce the accuracy of their aim. The balloon cables offered to snag their wings. When the wind was too strong for large balloons of the observation type, smaller, spherical ones were used up to heights of over 7000 feet. Their cables were often linked by transverse wires that formed an apron to entangle British bombers.

As with all Trenchard's creations, the Independent Force was efficiently organised and administered and highly effective in operation. The force flew 650 raids. Fifteen pilots and 35 observers/aerial gunners were killed: missing, respectively, amounted to 138 and 160; wounded, 39 and 34; injured 11 and 17.

The United States Air Service's late arrival and small numbers were compensated by the same enthusiasm, fighting spirit and bravery that the RFC, l'Aviation Militaire and the Italian Air Corps – and the Luftstreitkräfte – had shown.

In the short time at their disposal, the most successful pilots did prodigiously well. Captain E. V. Rickenbacker scored twenty-six victories, Second Lieutenant Frank Luke Jr twenty-one, First Lieutenant George A. Vaughn thirteen, Captain Field E. Kindley twelve.

Others, who served with the RFC or RNAS, included S. W. Rosevear with twenty-three, William C. Lambert twenty-four, Frederick W. Gillette, J. J. Malone, Alan M. Wilkinson twenty, Major Raoul Lufbery seventeen.

Hartney's 27 Squadron comprised one theatre owner, four salesmen, three lawyers, two journalists, five electrical and petroleum engineers, one concert pianist, one banker, one cotton planter, one motor racing driver, one mining man. As Hartney pointed out, all these jobs require clear, independent, constant thinking.

On 21st August, Major Hartney was given command of the 1st Pursuit Group, which comprised the 27th, 147th, 94th and 95th Squadrons. By the end of the war, the Group had destroyed 201 aircraft and balloons and lost 72 pilots dead, missing and wounded.

It was Hartney who devised a stratagem that became commonplace in the Second World War. Landing aircraft at night by the light of flares invited the now very active German bomber force to attack the aerodrome. Hartney had a dummy flarepath laid ten miles away. This attracted any enemy bombers in the vicinity and also acted as a beacon for his homing aircraft. Each pilot set course for base by this landmark, then flashed his individual letter by lamp shortly before entering the circuit. Three Aldis signal lamps were then shone across the airfield on the patch of ground where he should touch down; and extinguished as soon as he did. This was a hazardous way of landing, but much preferable to several sticks of bombs.

The total number of victories attributed to members of the USAS was 756 aeroplanes and 76 balloons. The method of awarding credits for victories to individuals, however, produced the anomalous total of 1513: because, whereas in the British and French Services, shared victory was divided between the participants who had fired at and hit the downed aeroplane or balloon, the Americans allowed one whole victory to everyone involved. In fact, 491 victories were each credited to one pilot; but, for the remaining 341, 1022 awards were handed out.

In addition to the fighter squadrons, there were an Observation Group consisting of three two-seater Spad squadrons, and a bombardment squadron, the 96th, all in operation by June 1918.

It must be remembered that not only were American squadrons in action over barely six months, but also the quality of the enemy aircraft and crews they fought was higher than it had been three years earlier. There were no easy victories now over obsolescent aeroplanes. Most of Rickenbacker's victims, for instance, were Albatroses. Pilots now reached the Front after thorough training.

Two of the most interesting American pilots were Douglas Campbell and Frank Luke Jr, contrasting types in origin and nature. Campbell's father was head of the Lick Observatory in California and sent Douglas to a private boarding school and Harvard. Luke was one of nine children who grew up in the Wild West pioneering environment of Montana, and saw more than one battle against Red Indians.

Campbell was twenty-two when he reached the 94th at the Western Front after virtually teaching himself to fly. Posted to the flying school at Issoudun as Adjutant, he had been fully occupied with office duties. He persuaded instructors to explain to him how to fly an aeroplane, and, without going through any dual, took off for the very first time in a Nieuport. He was the first American to shoot down five enemy machines; and those within six weeks. He scored six kills before he was grounded by a severe wound from an explosive bullet.

Luke was a simple, direct ranch hand whose honesty and lack of affectation often made his manner seem brash to his comrades. He was chatting to a sergeant one day when an enemy aircraft flew over the aerodrome. "Gee, that plane would be a cinch for me," he said. The sergeant denigrated this as boastfulness. It was not: Luke was merely making a naïf statement of self-confidence. What he really meant was that the aircraft would have been an easy target for anyone who happened to be airborne near it. He was not implying that he was the exceptional pilot who could have shot it down whereas others would have missed.

Despite his excellent record there were many who disliked him. Most of his comrades were gregarious, but he was shy, which prevented them from getting to know him well. He had one close friend, Lieutenant Joe Wehner, whose death in action caused him inconsolable grief. They had been attacking balloons when enemy fighters jumped them and Luke saw his crony shot down; but not before they had brought down two balloons and three enemy machines between them. Luke's comment to Hartney immediately after the event revealed a sensitive and considerate nature that was probably a great deal superior to his detractors'. "I'm glad it wasn't me. My mother doesn't know I'm on the Front yet." A boastful man would have lost no time in telling his mother he was in the fighting instead of concealing it to spare her feelings.

Luke's death was heroic, its aftermath tragic. The report of the Graves Registration Officer who found his grave on 29th September 1918 arouses deep sadness even seventy years on. "From the inspection of the grave and interviews held with the inhabitants of this town, the following information was learned with regard to this aviator and his heroism.

"Previous to being killed he had brought down three German balloons, two German aeroplanes and dropped hand bombs killing eleven German soldiers and wounding a number of others.

"He was wounded himself in the shoulder and evidently had to make a forced landing. Upon landing he opened fire with his automatic and fought until he was killed.

"It is also reported that the Germans took his shoes, leggings and money, leaving his grave unmarked."

The Allied offensive launched on 8th August 1918 swept the Germans back to ultimate surrender and an Armistice at 11 a.m. on 11th November. On the opening day, 1800 British, French and American aeroplanes swarmed over the enemy lines and beyond. The flying effort continued unabated until the 10th and great reputations were being enhanced even in the closing weeks of the war.

On 8th August, Beauchamp-Proctor, given the honour of leading his squadron, shared in the destruction of nine observation balloons, a record for a single day. On his 24th birthday, 4th September, he led a flight of five SE5as escorting RE8s. When the latter completed their bombing, the fighters carried on. His report said: "I saw some of our infantry advancing and a Hun machinegun team waiting to ambush them in a sunken road. I immediately dived and my flight followed. After two dives I could not see any of the machinegun crew alive. I flew round our infantry at twenty feet and saw them wave, then point towards some trenches. I climbed, then dived, whereupon thirty Huns attempted to get out of the trenches. They had been hiding in a dugout. Lt Corse, USAS, Capt. Carruthers and myself continued to shoot into the Huns until there were about five left and these were being engaged by our infantry. We engaged several small machinegun nests."

On 8th October he shot down a two-seater, then, single-handed, attacked eight German fighters. He was wounded in the shoulder but escaped by spinning. He was sent to hospital in England. Sholto Douglas recommended him for the VC in recognition of the zest he had shown for attacking the enemy in all circumstances. "For all his size, that little man had the guts of a lion," Major Douglas said of him.

On 27th October Major William Barker, DSO, MC and bar, who had forty-seven victories to his name, performed the most brilliant feat of arms achieved by a lone fighter pilot. Flying a Sopwith Snipe, he overtook a two-seater Rumpler at 21,000 feet and opened fire. The enemy observer's return fire destroyed Barker's telescopic sight. Using the peep sights, Barker hit the Rumpler's fuel tank, which caught fire. Before it exploded one of the crew parachuted to safety.

As he broke away he came under fire from a Fokker Triplane beneath. A bullet hit him in the right thigh. In great pain and with an almost useless right leg, he nearly fainted and the Snipe began to spin. When he recovered his senses and corrected his aircraft's attitude, he saw the whole of Geschwader 3's Jastas 2, 26, 27 and 36, totalling more than fifty Fokker D7s, above, beneath and around him. The nearest Jasta, twelve or fifteen strong, attacked him. He drove off two of them and shot one down in flames; but was himself hit in the left thigh. Loss of blood brought him to the verge of unconsciousness again. He went into a spin once more, came out of it and found himself in the midst of another Jasta. He brought one enemy down, was hit in the left elbow, fainted and spun down to 12,000 feet before regaining consciousness. Smoke was belching from his aircraft and he thought it must be on fire. He determined to ram the nearest Fokker, decided to chance a shot instead, and sent it down in flames. His engine was no longer emitting smoke: bullets had riddled one fuel tank and drained it. Intermittently insensible as he spun earthwards, unable to use the rudder bar and controlling the Snipe with stick and throttle only, he contrived to switch on the auxiliary tank and left his enemies behind. The sortie had lasted forty minutes. He crossed the British trenches at a height of a few feet, crashed, and the aeroplane turned over. Scottish soldiers extricated him before he bled to death. In hospital, he spent ten days in a coma. His legs healed in time for him to walk into Buckingham Palace to receive the Victoria Cross.

A French fighter pilot summed up the typical attitude of operationally experienced airmen, in *La Guerre Aérienne*: "Between missions, life was comfortable and enabled one to forget the war, but there was above all a great thirst to live life to the full during these short periods, in view of the uncertainties of the morrow; which they accepted. On operations, violent physical and nervous tension, caused by the fear of death."

However, after the first exchange of fire, fear disappeared. All energy was concentrated on a single purpose. "Opening fire is, for a pilot, a sort of mild doping, visions of agony become blurred and one surmounts them."

Another wrote: "The second time in action the shock is not so great. I had been warned. And then, of the number of shots fired, only five hit my aircraft, in harmless places. So one could get away with it and this was enough to give me the courage to carry on."

A bomber pilot put it: "Very quickly one armours oneself with indifference and recalls of these sorties more the irritants such as a too tight necktie, a rattle or vibration, than the brief moments of action."

A French fighter pilot with the unlikely name of Partridge shrugged off danger: "To fly alone, to be wounded, it's not a big deal and one accepts it as one of those inevitable things."

Even the most mettlesome sort of man, highly strung as a racehorse, the representative fighter pilot type, learned by experience to restrain his natural excitability and become phlegmatic in the cockpit.

Every man who took to the air in that first great air war must have been endowed with unusual courage: when every facet of air fighting had to be discovered for the first time; and the very machines in which they flew were so accident prone that merely being airborne was almost as dangerous as being under fire. Whatever their nationality, whether they were fighter pilots, who attracted the widest publicity, or aerial gunners who received none, courage was their hallmark, above skill at the controls or accuracy behind a gun.

Allies and enemy shared the same emotions when it was all over. Rickenbacker's description of the reaction of the 94th Squadron, which he commanded, on being told of the Armistice, expresses the feelings of all the airmen of all the nations that had been at war.

He received the news by telephone the evening before and read it out to his pilots. In the midst of a sudden silence a nearby artillery battery fired a joyous salvo. The pilots, shouting and tumbling over each other, rushed to their quarters and emerged firing pistols and Vérey lights into the air. Machineguns lit the darkness with tracer around the aerodrome. Searchlights swept the sky. Drums of petrol were rolled out of a hangar and set alight. Everyone joined hands and circled round the blaze, dancing crazily.

"I've lived through the war," someone yelled.

And somebody else shouted what most concerned them all: "We won't be shot at any more."

Aircraft Performance

To dogmatise about such data would be equivocal. Maximum speed at various heights, and rate of climb, of individual aeroplanes of the same type and with the same engine could differ widely. The tension of the rigging varied between individual machines. The general condition of engines, and their tuning, was not constant. The weight of pilot or crew was another variable. Aircraft instruments were not as accurate as they are now.

These facts put in doubt the validity of quoting speed in decimals or rate of climb in fractions of a minute. Where the best authority on each aeroplane shown below does so, these are shown here to the nearer mile per hour or minute respectively.

Comparison between the speeds of one make and mark and another is invidious, because the parameters are not consistent. In most instances, speed at sea level is stated; but the heights above that for which maximum speeds are given vary. So do the heights to which climb was timed.

Where there are divergences in performance of an aeroplane that was fitted with more than one make or horse power of engine, the higher or highest is given here.

Aircraft	Engine	Max. Speed	Max. Speed at 10,000 ft	Climb Time	Height
Shorthorn	Renault 80 h.p.	72 m.p.h.	72 m.p.h.	8 mins	500m
Farman F20	Gnome or Le Rhône 80 h.p.	65 m.p.h.		8 mins	500m
Martinsyde S.1	Gnome 80 h.p.	87 m.p.h.			
Voisin LA	Salmson 130 h.p.	65 m.p.h.		30 mins	2000m
Caudron G3	Gnome 80 h.p.	65 m.p.h.		27 mins	2000m
BE2c	RAF 90 h.p.	72 m.p.h.	69 m.p.h.	45 mins	
Fokker E1	80 h.p.	81 m.p.h.		40 mins	10,000 ft
Morane-Saulnier N	Le Rhône 80 h.p.	90 m.p.h.			
DH2	Gnome 100 h.p.	93 m.p.h.		25 mins	10,000 ft
FE2b	Beardmore 120 h.p.	80 m.p.h.	72 m.p.h.	52 mins	10,000 ft
FE 8	Gnome 110 h.p.	at 6000 ft 79 m.p.h.	69 m.p.h.	24 mins	10,000 ft
RE8	RAF 140 h.p.		92 m.p.h.	21 mins	10,000 ft
Nieuport 11	Gnome 80 h.p.	97 m.p.h.		16 mins	2000m

Sopwith 1½-Strutter	Clerget 110 h.p.	105 m.p.h.	96 m.p.h.	21 mins	10,000 ft
Sopwith Pup	Le Rhône 80 h.p.	111 m.p.h.	102 m.p.h.	17 mins	10,000 ft
FE8	Gnome 110 h.p.		69 m.p.h.	24 mins	10,000 ft
Caproni Ca3	Three Isotta-Fraschini 190 h.p.	85 m.p.h.		13 mins	2000m
Sopwith Triplane	Clerget 130 h.p.	114 m.p.h.	106 m.p.h.	12 mins	10,000 ft
Albatros D3	Mercedes 160 h.p.	110 m.p.h.	103 m.p.h.	14 mins	10,000 ft
Spad 7	Hispano-Suiza 150 h.p.	120 m.p.h.	116 m.p.h.	12 mins	10,000 ft
DH4	Rolls-Royce Eagle 3 250 h.p.		113 m.p.h.	17 mins	10,000 ft
Bristol Fighter F2b	Rolls-Royce Falcon 3 275 h.p.		113 m.p.h.	11 mins	10,000 ft
SE5a	Wolseley 200 h.p.		117 m.p.h.		
Sopwith Camel	Clerget 130 h.p.		104 m.p.h.	12 mins	10,000 ft
Handley Page 0/100	Two Rolls-Royce Eagle 250 h.p.	95 m.p.h.			
Pfalz D3	Mercedes 160 h.p.		102 m.p.h.	41 mins	15,000 ft
DH9	Siddeley Puma 230 h.p.		112 m.p.h.	21 mins	10,000 ft
Nieuport 17	Le Rhône 110 h.p.	at 3000m 103 m.p.h.	96 m.p.h.	12 mins	3000m
Spad 13	Hispano-Suiza 220 h.p.	at 3000m 133 m.p.h.		8 mins	3000m
Fokker D7	Mercedes 160 h.p.	130 m.p.h.	124 m.p.h.	10 mins	10,000 ft
Sopwith Dolphin	Wolseley 200 h.p.		128 m.p.h.	11 mins	10,000 ft

Bibliography

Baring, Maurice, *Flying Corps Headquarters 1914–1918*, Bell, 1920
Bartlett, C. P. O., *Bomber Pilot*, I. Allan, 1974
Bishop, Lt. Col. William A., *Winged Warfare*, Bailey Brothers & Swinfen, 1975
Boyle, Andrew, *Trenchard, Man of Vision*, Collins, 1962
Bradshaw, Stanley Orton, *Flying Memories*, Hamilton
Clark, Alan, *Aces High*, Collins, 1973
Collishaw, R., *Air Command*, Kimber, 1973
Crundall, E. D., *Fighter Pilot on the Western Front*, Kimber, 1975
Degelow, Carl, *Germany's Last Knight of the Air. The Memoirs of Carl Degelow*, Kimber, 1979
Douglas, Sholto, *Years of Combat*, Collins, 1963
Dudgeon, James M., *Mannock, VC, DSO, MC, RFC, RAF*, Hale, 1981
Editors of *Life* Magazine, *The First World War*, 1964
Funderbank, Thomas R., *The Early Birds of War*, Jarrold, 1973
Grider, John M., *War Birds. The Diary of an Unknown Aviator*, 1926
Grinnell-Milne, Duncan W., *Wind in the Wires*, Jarrold, 1971
Liddell Hart, Sir Basil, *History of the First World War*, Cassell, 1970
Hartney, Harold E., *Wings over France*, Bailey Brothers & Swinfen, 1974
Hawker, Tyrrel M., *Hawker, VC*, Mitre Press, 1965
Illingworth, Capt. A. E. and Robeson, Major V. A. H., MC, *A History of 24 Squadron RAF*
Immelmann, Franz, *Immelmann: The Eagle of Lille*, Hamilton, 1935
Insall, Algernon J., *Observer: Memories of the RFC*, Kimber, 1970

Jones, H. A., *The War in the Air*, Vols 2–5, HMSO & Hamish Hamilton, 1969

Lee, Arthur Gould, *No Parachute*, Jarrold, 1967

Lewis, Cecil, *Sagittarius Rising*, Peter David, 1936

McKee, Alexander, *The Friendless Sky*, Elmfield Press, 1973

Macmillan, Norman, *Offensive Patrol. The Story of the RNAS, RFC, RAF in Italy 1917–1918*, Jarrold, 1973

McScotch, *Fighter Pilot*, Routledge, 1936

Montgomery-Moore, Cecil, *That's My Bloody Plane*, Peuquot, 1975

Morris, Alan, *First of the Many*, Jarrold, 1968

Munson, Kenneth, *Bombers 1914–19*, Blandford Press, 1968

Neumann, Georg P., *German Air Force in the Great War*, Chivers, 1969

Noble, Walter, *With a Bristol Fighter Squadron*, Chivers, 1975

Pilot, by a, *War Flying*, John Murray, 1916

Raleigh, Sir Walter, *The War in the Air*, Vol. 1, HMSO & Hamish Hamilton, 1969

Rickenbacker, Captain Eddie V., *Fighting the Flying Circus*, Bailey Brothers & Swinfen, 1973

Schroeder, Hans, *An Airman Remembers*, Hamilton, 1935

Shaw, Michael, *Twice Vertical, The History of No. 1 Squadron, RAF*, Macdonald, 1971

Springs, Elliot White, *War Birds*, Temple Press, 1966

Titler, Dale M., *The Day the Red Baron Died*, Ian Allan, 1973

Tredrey, F. D., *Pioneer Pilot. The Great Smith Barry Who Taught the World How to Fly*, Peter Davies, 1976

Warplanes of World War I, Phoebus, 1973

Whitehouse, Arch, *The Fledgling*, Vane, 1965

Winter, Denis, *The First of the Few*, Allen Lane, 1982

Woodhouse, Jack, and Embleton, G. A., *The War in the Air 1914–1918*, Almark, 1974

Yeates, V. M., *Winged Victory*, Cape, 1934

Porret, D., *Les "AS" Français de la Grande Guerre*, Cedocar, 1983

La Grande Guerre dans le Ciel, Centre Historique de l'Armée de l'Air

La Guerre Aérienne, several vols, Centre Historique de l'Armée de l'Air

Corsini, Paolo, *La Partecipazione degli Aviatori Inglesi alle Operazione sul Fronte Italiano durante la Grande Guerra e il Cinquantenario della Fondazione della RAF*, Ufficio Storico Aeronautica Militare

Ludovici, Domenico, *Gli Aviatori Italiani del Bombardamento nella Guerra 1915–1918*, Stato Maggiore Aeronautica, Ufficio Storico, Roma, 1980

Mencarelli, Igino, *Fulco Ruffo di Calabria*, Ufficio Storico Aeronautica Militare, 1970

Mencarelli, Igino, *Pier Ruggero Piccio*, Ufficio Storico Aeronautica Militare, 1970

Mencarelli, Igino, *Francesco Baracca*, Ufficio Storico Aeronautica Militare, 1969

Porro, Generale Sq. A. Felice, *La Guerra nell'Aria 1915–1918*, Edizioni Mate, Milano, 1965

Porro, Generale Alberto, *Natale Palli*, Stato Maggiore Aeronautica Militare, Ufficio Storico, Roma, 1973

Silvio Scaroni, Ufficio Storico Aeronautica Militare, 1969

Istruzione di Allievi Piloti Americani nella Prima Guerra Mondiale, Stato Maggiore Aeronautica Militare, Ufficio Storico, Roma, 1956

Arndt, Major A.D. Hans, *Die Fliegertrupper im Weltkriege*, Archivrat am Reichsarchiv

Gener, H., *Deutschlands Luftfahrt und Luftwaffe*, Walter de Grunter & Co., Berlin, 1937

Gron, Hermann, *Die Organisation des Deutchen Heeres im Weltkriege*, E. S. Mittler & Sohn, Berlin, 1923

Haupt-Heydemarck, A. D., *Flieger im Westen*, Wilhelm Rolf, Berlin, 1931

von Loewenstern, Baron Elard, *Der Frontflieger*, Berlin, 1937

Zuerl, Walter, *Pour le Mérite-Flieger*, Curt Pechstein, München, 1938

Index

Balloons: First hot air, 6. First hydrogen, 6.
First British Army Balloon Unit
formed. At reviews and manœuvres. On
active service in Bechuanaland and
Sudan. Photography from free balloons
7, 8. In Germany, 8. In Italy, 8
Balloon Factory, H. M.: 10
Balsley, Clive: Pilot, Escadrille Lafayette, 116
Bannerman, Lieutenant Colonel Sir
Alexander: First Commanding Officer
of Air Battalion, 16
Baracca, Tenente Colonnello Francesco:
Training in France and joins squadron,
85. First three combats, 86. First
victory, 133. Second victory, 134, 135.
Commanding 91st Sqdn, 224. Wins
Gold Medal, 233. Killed, 251
Barbieri, Tenente Colonnello Alfredo: 160, 16
Barès, Lieutenant Colonel Joseph: Becomes
Responsable de l'aéronautique for the
battle area, 67. Character, 69. Use of
strategic bombing. Use of fighters, 76.
At Verdun, 110
Baring, Major The Hon. Maurice: 29, 30, 35,
38, 39, 45, 46, 101, 206
Barker, Major William: On 28 Sqdn, 231.
Attacks Austro-Hungarian aerodrome,
232. Score equalled by Mannock, 242.
Commanding 139 Sqdn, 252. Shoots
down four in one flight, 258
Barrachini, Tenente Flavio: 225, 250
Barrington-Kennett, Lieutenant B. H.: 23
Bayly, Lieutenant, C. G. G.: First RFC
observer shot down behind German
lines and taken prisoner, 35
Baynes, Lieutenant: 244
Beardmore Engines: 4, 118, 165
Beauchamp-Proctor, Captain A. W.:
Introduced, 226, 227. Early victories,
239. On 84 Sqdn, 248, 257
Bell Davies, Wing Commander: 146
Berliner Tageblatt: On Italo-Turkish War, 14
Bernard, General: Air Force Commander at
French General Headquarters 1914, 67
Beuermann, Professor: Balloon pilot, 59
Bishop, Major William A. ("Billy"):
Introduction to, 194. Early career, 195.
Encounter with "Flying Pig", 196.
Attacks balloon, 197. Hatred of
Germans, 201. Earns VC by solo attack
on aerodrome, 202. Describes William
Barker as world's best fighter pilot, 231.
Succeeded by Mannock as CO 85
Sqdn, 244. Leads formation of 19 SE5as
to France, 245
Blériot, Louis: 12, 63

Blomfield, Major R. G.: Commanded No. 56
Sqdn and assembled a squadron
orchestra, 189, 210, 211
Blondeau, Major R. G.: Partner in a Brooklands
flying school, 13
Blue Max: *see* Pour le Mérite
Bluebird Restaurant: Brooklands, Social
Centre, 13
Boelcke, Hauptmann Oswald: On Flying
Section No. 62, 58. Proving Fokker E1,
96. Defines fighter tactics with
Immelmann, 98. At Douai, 99. At
Rethel to innovate barrage patrols, 102.
Combat with Sholto Douglas, 106, 107.
Eighth victory. Awarded Pour le
Mérite, 108. Solo sorties from forward
airstrip, 111. Trouble with interrupter
gear, 111. Flying Fokker E4, 111.
Grounded by Kaiser. Commands
Jagdstaffel 2, 137. Jasta 2 on the Somme,
143. Killed, 143, 144, 145. Record
passed by Richthofen, 186
Böhme, Leutnant Erwin: 144
Bologna, Tenente Luigi: 161, 162
Borton, Lieutenant A. E.: Coined term
"Archie" for anti-aircraft fire, 41, 69
Bourdillon, 2nd Lieutenant R. B.: With Louis
Strange, designed early bombsight, 73
Bowles, Lieutenant F. S.: 252
Bowman, Lieutenant: 210
Bozon-Verduraz, Sous Lieutenant Louis: 214
Brancker, Major General W. Sefton: Cavalry
Manœuvres in India. Learns to fly and
attends CFS. Deputy Director of
Military Aeronautics, 33. Discusses
Henderson with Kitchener, 45.
Commanding a wing, 103
Brocard, Capitaine Félix: 110, 111
Brooke-Popham, Lieutenant Colonel: Opinion
of Henderson, 19. Commands No. 3
Squadron, 23. Makes first aeroplane
reconnaissance, at 1913 manœuvres,
24. HQ RFC France, 94, 96
Brooklands: Temporary aerodromes prepared
for Louis Paulhan's display. Permanent
aerodrome laid out. Becomes cradle of
British aviation, 13. Flying Schools, 17,
18, 21, 91, 102
Brooks, Lieutenant: 151
Brown, Lieutenant Roy: Wrongly credited with
having shot down Manfred von
Richthofen, 253
Buchan, John: Tribute to Henderson, 18
Buie, Gunner Robert: Proved to have shot
down Manfred von Richthofen, 253
Bulgaria: 21, 115

Magistrini, Sergente Cesare: 225
Magoun, Lieutenant: 244
Maitland, Captain Edward: Commands No. 1
    Sqdn, 23
Malcolm, Major G. J.: 118, 164, 165, 166
Malone, J. J.: Among leading American pilots,
    255
Mannock, Major Edward: Most successful
    RFC pilot. Early life. Goes to Turkey.
    Interned. Joins RAMC, 88, 89, 90.
    Joins RE as officer cadet. Applies for
    RFC. Qualifies as pilot, 179. On No. 40
    Sqdn. Hysteria in action, 193. Record,
    194. Discouraged, 203. First victory,
    204. Encourages McLanachan, 205.
    Hatred of incendiary ammunition, 206.
    Instructing in England. Joins No. 74
    Sqdn, 240. Mentally unbalanced
    behaviour, 241. Commands No. 85
    Sqdn, 241. Leads aerial ambush. Effect
    of McCudden's death on him. Killed,
    242. No. 85 Sqdn had requested him as
    CO, 245
Mansfield, Major W. H. C.: 37, 166
Mapplebeck, Lieutenant G. W.: Flies RFC's
    first operational sortie, with Joubert de
    la Ferté, 34, 35. Killed, 73
Mateki Darevni: 23
Maubeuge: RFC Expeditionary Force
    assembles at, 3, 34
May, Lieutenant Wilfred: 253
Mayfly: Royal Navy's first airship, 9
McConnell, James: 115, 125
McCubbin, Second Lieutenant: 137
McCudden, Major James: Mechanic on No. 3
    Squadron, 24. On 1914 retreat, 36.
    Flies as observer, 43. Discomforts of
    active service, 52. Days at forward
    aerodrome, Fosse, 53. Passes navigation
    test for trainee observer. Promoted to
    corporal, 55. Sergeant mechanic, flying
    as observer and awaiting pilot's course.
    Comments on Ludlow-Hewitt, 90.
    With Harvey-Kelly as pilot, combat
    with Immelmann, 91. Qualifies as pilot
    and Joins No. 20 Sqdn. Posted to No.
    29 Sqdn. Commissioned, 178. Meets
    Mannock, 179. Target practice, 193.
    Instructing at CFS, 210. Joins 56 Sqdn,
    211
McCudden, Sergeant W. T. M.: Pilot, 24
McElvain, Lieutenant: 248
McKay, Lieutenant: 143
McLanachan, Lieutenant: 204, 205
Merville: 54
Messimy: French Minister of War, 67

Military Aeronautical Directorate: 24
Military Training, Director of: 29
Military Wing, RFC: Function of, 22.
    Trenchard takes command of, 31
Miller, Captain James: Commanded 95th
    Pursuit Squadron, 243
Miller, Major A. M.: Recruiting tour of South
    Africa, 227
Millerand: French Minister of War, 67
Miraglia, Giuseppe: 64, 87
Mons: 36, 37
Montenegro: 22
Montgolfier, Jacques: 6
Montgolfier, Joseph: 6
Montrose: 25, 26
Moore-Brabazon, Lieutenant: 72
Moris, Colonnello M. M.: Director General,
    Italian Military Air Corps, 84
Morris, Lieutenant, L. F. B.: 143
Mortane, Jacques: 126
Moulinais, Brindejonc des: 124, 125, 126
Mulock, Squadron Commander Redford: 190

N

Nardini, Sergente Guido: 225
Nash, Flight Sub-Lieutenant Gerry: 191
Naval Wing: Its function not yet fully defined,
    23
Navarre, Captaine Jean: Early career, 56.
    Transfers from bombers to fighters,
    57. First encounter with enemy. At
    Verdun. Tactics. Shot down. Post-war
    flying and death, 112, 113
Neuve Chapelle, Battle of: 72, 75
Newall, Lieutenant Colonel Cyril: 149, 254
Nivelle, General: 213
Nixon, Captain: 193
Noel, Lieutenant N.: 35
Noel, Louis: Eccentric flying instructor.
    Taught Strange, 55
Northcliffe, Lord: 237
Novelli, Tenente Gastone: 225
Nungesser, Capitaine Charles: Early career.
    Transfers from bombers to fighters, 54,
    55, 56. Personal markings on aeroplane,
    112. Frequent crashes, 113. Fights six
    Albatroses, 214. Crashes car, 215

O

Oberursel Engine: 96, 97, 111
Observers: No official category or training, 42.
    Observers' badge instituted: 51
O'Connell, Lieutenant J. Rutledge: 251
Olivari, Tenente Luigi: 135, 225

[273]

Omdurman, Battle of: Henderson at, 28
Osnaghi, Tenente: 251
Ostend: 3

P

Paardeburg: 9
Paine, Captain Godfrey, R. N.: Commandant, Central Flying School, 22
Parabellum Machinegun: 103, 123, 188
Parachutes: 1, 23, 122, 175, 176, 177
Paris: First balloon ascent from, 6. Enemy approaching, 38. Germans bomb, 42.
Parker, Second Lieutenant J.: 97
Parschau, Leutnant: 96
Partridge, Lieutenant: 259
Paulhan, Louis: 12
Paynter, Mrs: 172
Peck, Second Lieutenant R. J.: 97
Pégoud, Adolphe: 98, 122, 123, 175
Pell, Lieutenant: 192
Perry, Mr: 21
Perry, Lieutenant E. W. C.: 34
Pétain, Marshal: 110, 213
Peterson, Captain: On US Air Service's first operational patrol, 243, 244
Phillips, Lieutenant G. E.: 8
Photography: From small free balloons, 8. First operational photographic reconnaissance, 41. Early difficulties from aeroplanes, 51. Photographic Sections set up at all Wing HQs, 72. Routine, 149
Piave River, Battle of: 249
Piazza, Capitano Carlo: 14
Piccio, Tenente Colonnello: Commander of First Italian Fighter Group, 225, 237, 251
Pisa: Flying school, 86
Pixton, Captain Howard: 18
Playfair, Captain: 103
Plaz: Fokker's Chief Designer, 219
Pochon, Private: 56
Point Blank: Definition of, 113
Portal, Lieutenant Arthur: 54
Poser, Hauptmann von: 83
Pour le Mérite: 60, 123
Pourpe, Marc: Pioneer French civilian and military pilot. Lufbery's employer, 49
Pretyman, Lieutenant G. F.: Flies RFC's first photographic reconnaissance, 41. Experimental night flying, 53
Prince, Norman: With Dr Gros, founds *Escadrille Lafayette*, 115, 116
Purdey, Captain: 179
Puttnam, Lieutenant: 244

Q

Quenault, Louis: Observer in first aircraft to shoot down an enemy aeroplane, 42

R

Raleigh, Major G. H.: 24
Ranza, Tenente Ferrucio: 225
Raymond, Lieutenant: 240
Rees, Lieutenant T.: 143
Reid, Flight Sub-Lieutenant Ellis: 191
Reims: 14, 48, 85
Renault Engines: 4
Rhodes-Moorhouse, Lieutenant: Wins RFC's first VC, 74
Rhône, Le, Engines: 14
Richthofen, Hauptmann Lothar von: 193, 194
Richthofen, Ritter Manfred von: Most successful fighter pilot of the war. Meets Boelcke, 137. Joins Jasta 2, 138. On the Somme, 143. Shoots down Hawker, 144, 145. Attitude to crossing the lines, 154. Report on Sopwith Pup, 178. Leads attack on No. 40 Sqdn, 180. Passes Boelcke's record, 186. First combat with Bristol Fighters, 188. Encounters with Naval Three, 190. Opinion of Sopwith Triplane, 191. Bombed by RFC, 193. His Jasta in combat with Ball's squadron, 194. Lust for killing, 201. Some of his pilots fight Bishop, 202. Character, 203. Commands first *Jagdgeschwader*, 217, 218. Wounded, 219. Resumes operations, 229. Leads big mixed formation, 238. Killed, 253.
Rickenbacker, Major Edward: Introduced, 49. Commissioned in US Air Service. Joins 94th Sqdn, 242. On US Air Service's first operational patrol. First Victory, 244, 245. Total score, 255. Assessment of ability, 256. Reaction to news of armistice, 259.
Ridd, pilot Corporal F.: 24
Ridley, Second Lieutenant C. A.: 90. Forced landing behind enemy lines and escape, 152, 153
Rockwell, Kiffin: 115, 116, 171
Rodney, Second Lieutenant the Hon. W. F.: 75
Roe, A. V.: 12
Rolls-Royce Engines: 4, 165, 184, 228
Rosevear, S. W.: Among leading American pilots, 255
Röth, Ritter Fritz von: 60, 61, 219, 220, 252
Rothermere, Lord: 237
Rougier, Henri: Pioneer French pilot, 63

P. R. REID

COLDITZ

For the first time in one paperback volume the legendary
story of Colditz.

Oflag IVC was the most dreaded of all the German
P.O.W. camps of World War II – with the reputation of
being totally escape-proof.

This is the story of those men who refused to concede
defeat, the indestructible prisoners-of-war who, against
all the odds, managed to escape and make their way to
freedom. By ingenuity, cunning and sheer daredevil
bravado they achieved the impossible – and their heroism has rightly passed into the annals of history . . .

A Royal Mail service in association with the Book Marketing Council & The Booksellers Association.
Post-A-Book is a Post Office trademark.

ION PACEPA

RED HORIZONS
THE EXTRAORDINARY MEMOIRS OF A
COMMUNIST SPY CHIEF

Lieutenant General Ion Mihai Pacepa was no run-of-the-mill defector.

Head of Rumanian Intelligence and a personal advisor to President Ceausescu, he knew everything that went on among the ruling elite – so much so that his debriefing took three years.

Red Horizons is his startling, behind-the-scenes description of a country, in name Communist but in fact a brutal dictatorship, where corruption, grotesque drunkenness, nepotism, personal greed and brutality rule . . .

'The president is seen ordering assassinations, plotting to steal technology, and periodically dancing with anger when things go wrong'

The Times

He describes with horrifying candour what is perhaps the most corrupt tyranny surviving in Europe'
    C. M. Woodhouse, *The Times Literary Supplement*

HODDER AND STOUGHTON PAPERBAKCS

ANN & JOHN TUSA

THE BERLIN BLOCKADE

A major city supplied entirely by air for over a year.

That was the unique, historic achievement of the airlift that broke the Berlin Blockade.

Everything, not just food but coal, petrol, the whole staggering range and weight of supplies for well over 2 million people, had to be flown in.

Day in, day out, month after month, the transport planes, British and American, military and civil, flew in. A flight every ninety seconds was the average. Turn round time came down to six minutes. The strain on aircraft, crew and organisation was almost unbearable at times. But it was done.

Berlin, still largely war-ruined, its people, cold, underfed and apprehensive, was saved. The Russians accepted defeat. The blockade was lifted.

A political turning point and a milestone in the history of flying, the Berlin Blockade is also a fascinating story.

'Painstakingly researched and eminently readable'
The Sunday Times

HODDER AND STOUGHTON PAPERBACKS

CORNELIUS RYAN

A BRIDGE TOO FAR

'Unquestionably the most brilliant account of a battle that I have ever read'
> Lt. General Sir Brian Horrocks, *The Spectator*

'Tremendous, towering, magnificent ... a moving, awesome and accurate portrayal of human courage'
> General James M. Gavin

'A magnificently readable account of an operation that has always caught the public imagination'
> *Daily Telegraph*

'Quite markedly Mr. Ryan's best book'
> *The Times*

'Presented with absolute mastery of the situation. Magnificent'
> A. J. P. Taylor, *The Observer*

HODDER AND STOUGHTON PAPERBACKS

WINSTON S. CHURCHILL

THE RIVER WAR

In 1881 the Mahdi's rebellion plunged the Sudan into bloodshed and confusion. Egyptian armies sent to recover the territory were routed and destroyed. All outside control and administration had been wiped out. Mr Gladstone's Government decided that British interests in the area were to be withdrawn. General Gordon was sent to Khartoum to bring out the surviving officials, soldiers and Egyptian subjects. But, as the Mahdi's forces surrounded Khartoum, Gordon was trapped and doomed.

The River War tells of the expedition of reconquest that, under General Kitchener, fought its way up the Nile. The young Winston Churchill was there. This is his classic account of the expedition and the final Battle of Omdurman.

HODDER AND STOUGHTON PAPERBACKS

MORE TITLES AVAILABLE FROM
HODDER AND STOUGHTON PAPERBACKS

☐	38631 2	**P. R. REID** Colditz	£4.99
☐	49745 9	**ION PACEPA** Red Horizons: The Extraordinary Memoirs Of A Communist Spy Chief	£4.99
☐	50068 9	**ANN & JOHN TUSA** The Berlin Blockade	£4.99
☐	19941 5	**CORNELIUS RYAN** A Bridge Too Far	£4.99
☐	42213 0	**TREVOR ROYLE** The Best Years Of Their Lives	£3.95

All these books are available at your local bookshop or newsagent, or can be ordered direct from the publisher. Just tick the titles you want and fill in the form below.

Prices and availability subject to change without notice.

Hodder & Stoughton Paperbacks, P.O. Box 11, Falmouth, Cornwall.

Please send cheque or postal order, and allow the following for postage and packing:

U.K. – 55p for one book, plus 22p for the second book, and 14p for each additional book ordered up to a £1.75 maximum.

B.F.P.O. and EIRE – 55p for the first book, plus 22p for the second book, and 14p per copy for the next 7 books, 8p per book thereafter.

OTHER OVERSEAS CUSTOMERS – £1.00 for the first book, plus 25p per copy for each additional book.

Name...

Address..

....